MANAGING A SEA

MANAGING A SEA

THE ECOLOGICAL ECONOMICS OF THE BALTIC

edited by

Ing-Marie Gren, R Kerry Turner and Fredrik Wulff

BEIJER The Beijer International Institute of Ecological Economics
The Royal Swedish Academy of Sciences

EARTHSCAN
Earthscan Publications Ltd, London

First published in the UK in 2000 by
Earthscan Publications Ltd

Copyright © The Beijer Institute, 2000

A catalogue record for this book is available from the British Library

ISBN: 1 85383 608 7

The research project on which this book is based was funded by the European
Commission between 1992 and 1995 (project no EV5V-CT-92-0183). The opinions
expressed by the authors do not necessarily reflect those of the Commission

Typesetting by Composition & Design Service

Printed and bound by Creative Print and Design Wales, Ebbw Vale
Cover design by Yvonne Booth

For a full list of publications please contact:

Earthscan Publications Ltd
120 Pentonville Road
London, Nl 9JN, UK
Tel: +44(0) 171 278 0433
Fax: +44(0) 171 278 1142
Email: earthinfo@earthscan.co.uk
http://www.earthscan.co.uk

Earthscan is an editorially independent subsidiary of Kogan Page Ltd and publishes
in association with WWF-UK and the International Institute for Environment and
Development

This book is printed on elemental chlorine free paper

CONTENTS

PART III: INSTITUTIONS AND POLICIES

List of Figures and Tables

Figures

TABLES

About the Contributors

Carl Folke is an ecologist and professor in natural resource management at the Department of Systems Ecology, Stockholm University. His main research activities focus on the interactions between human and natural ecosystems.

Ing-Marie Gren is an economist and professor in natural resource and environmental economics at the Department of Economics, Swedish University of Agricultural Sciences, Uppsala. She specializes in the economics of water quality management with emphasis on the role of policy instruments and their enforcement.

Åsa Jansson is a doctoral student with a degree in natural resource management. Her doctoral work at the Department of Systems Ecology, Stockholm University, is focused on calculating the value of ecological services for human society.

Åsa Johannesson has a degree in economics from the Stockholm School of Economics. Her thesis work, together with Paula Randås, included the estimation of economy-wide costs for nitrogen reductions in a selection of Baltic Sea countries.

Agnieska Markowska is a doctoral student at the Department of Economics, Warsaw University. Her thesis work involves monetary valuation of water quality changes.

Tore Söderqvist has a doctorate in natural resource and environmental economics. His research at the Beijer International Institute of Ecological Economics focuses on the valuation of environmental changes in monetary terms.

Paula Randås has a degree in economics from the Stockholm School of Economics, Stockholm. Together with Åsa Johannesson she wrote her thesis on the comparisons of policies for nitrogen reductions in a selection of Baltic Sea countries.

Kerry Turner is an economist, professor in natural resource and environmental economics and Director of the Centre for Social and Economic Research on the Global Environment (CSERGE) at the University of East Anglia, Norwich. His main interests are monetary valuation of environmental changes and economics of waste management. He is author and co-author of several text-books in environmental and resource economics.

Fredrik Wulff is a systems ecologist and professor in marine biology at the Department of Systems Ecology, Stockholm University. He is currently modelling large scale dynamics of nutrient transports in the Baltic Sea, and investigating large scale impacts from various functioning of coastal zones.

Tomasz Zylicz is an economist and professor in economics at the Department of Economics, Warsaw University. He is interested in the design of environmental policies in general, and policies for improving the biological conditions of the Baltic Sea in particular.

ACRONYMS AND ABBREVIATIONS

C	carbon
CFC	chlorofluorocarbon
CGE	computable general equilibrium
CVM	contingent valuation method
DC	discrete choice
EC	European Commission
EU	European Union
GDP	gross domestic product
GIS	Geographical Information System
HELCOM	Helsinki Commission
JCEAP	Joint Comprehensive Environmental Action Programme
kg	kilogram
km	kilometre
MC	marginal cost
mM	millimole
N	nitrogen
N_2	nitrogen gas
NH_3	ammonia
NH_4	ammonium
NO_3	nitrate
NO_x	nitrogen oxides
NP	non-protesters
NR	non-respondents
OE	open-ended
OECD	Organisation for Economic Co-operation and Development
P	phosphorus
PLN	Polish zloty
P-S-I-R	pressure-state-impacts-response
SAM	Social Accounting Matrix
SEBR	South-east Baltic Region
SEK	Swedish krona
SP	strong protesters
TCM	travel cost method
TIN	total inorganic nitrogen
TN	total nitrogen
WP	weak protesters
WTP	willingness to pay
z	price of permits per unit nutrient reduction

1 INTRODUCTION

Ing-Marie Gren, Kerry Turner and Fredrik Wulff

The process of global environmental change manifests itself across a variety of scales, with the world's regional seas such as the Baltic, Black, North and Mediterranean Seas and, more generally, coastal zones being among a group of areas that are under particularly intense environmental pressure. These 'enclosed' seas and coastal areas are multiple-use zones in which a number of different stakeholders are looking to impose often conflicting resource demands – that may include fishing, recreation and amenity, transportation and trade, energy and minerals extraction and nature conservation – on a finite resource base. In many places, this human-generated environmental pressure is exacerbated by natural variability in biogeochemical processes and climate change, so that marine ecosystems and coastal zones have become more vulnerable and less resilient to further stress and shock.

A marked feature of the pollution and resource overexploitation problems in the regional seas and coastal zones is the significance of 'out-of-zone' activities and their effects. Most of the damage occurring within these places is related to activities located in the wider drainage basin and beyond. In the first wave of modern environmentalism in the 1960s the focus was directed at the reduction of point-source pollution through 'end-of-pipe' technologies. While considerable scope remains for further mitigation activity of this sort, particularly in the South-east Baltic Region (SEBR) (see Figure 1.1), the future challenge lies more with curbing diffuse sources of pollution caused by land-use change and transportation.

The cost to society of buffering the economic and environmental capital assets from the effects of change at or near shorelines has been escalating steadily. A political protection cycle is also stimulated in which further socio-economic development is encouraged to locate in potentially hazardous locations because of the expectation that protection will be maintained and/or enhanced. Expensive 'hard engineering' structures and works (for example, sea walls, barrages, beach groynes and near shore concrete reefs) litter many coastlines, but the power of 'natural' coastal forces is such that a major re-think is underway in which measures that work with nature ('soft engineering') are increasingly being implemented, for example sand dune protection, salt marsh protection and beach augmentation. Society will be asked

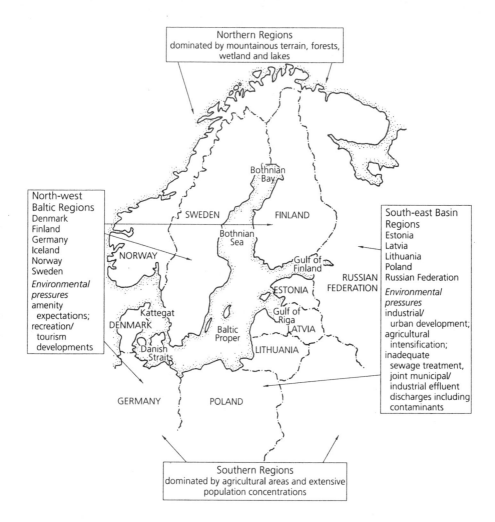

Northern Regions
dominated by mountainous terrain, forests, wetland and lakes

Bothnian Bay

North-west Baltic Regions
Denmark
Finland
Germany
Iceland
Norway
Sweden
Environmental pressures
amenity
  expectations;
recreation/
  tourism
  developments

SWEDEN

Bothnian Sea

FINLAND

Gulf of Finland

NORWAY

RUSSIAN FEDERATION

ESTONIA

Gulf of Riga

Kattegat
DENMARK

LATVIA

Baltic Proper

Danish Straits

LITHUANIA

South-east Basin Regions
Estonia
Latvia
Lithuania
Poland
Russian Federation
Environmental pressures
industrial/
  urban development;
agricultural
  intensification;
inadequate
  sewage treatment,
joint municipal/
industrial effluent
discharges including
contaminants

GERMANY   POLAND

Southern Regions
dominated by agricultural areas and extensive population concentrations

Figure 1.1 *The Baltic Drainage Basin – Watershed Regions*

to accept coastal realignment in many areas in the future, rather than sea defence and coastal protection works (Bower and Turner, 1998; Turner et al, 1998).

Marine pollution has, until recent decades, been considered as only a relatively localized problem, primarily due to the vast dilution effect of the oceans. The Baltic Sea was the first case where environmental changes to an entire marine ecosystem were documented. The virtual elimination of top predators such as eagles and seals, the development of anoxia in deep basins and toxic blooms are examples of large-scale environmental effects caused by the release and accumulation of toxic substances and the eutrophication related to the opening of nutrient cycles, giving nitrogen and phosphorus more mobility, in the Baltic.

It is, however, not surprising that the Baltic is particularly sensitive to pollution, given the loading imposed by about 85 million people combined with intensive agriculture and industry in the surrounding drainage basin, which is four times larger than the sea. The special physical properties of this enclosed shallow brackish-water sea further enhance the effects of the large inputs of polluting substances. Originally an oligotrophic system, a four- to eightfold increase of nitrogen and phosphorus inputs during this century has increased productivity and dramatically changed the species composition and food webs. New species have been established as important components of the Baltic. The ultimate purpose of this book is to identify strategies for an effective and efficient improvement in the biological condition of the Baltic Sea and the sustaining of such conditions in the future.

The content of the book is unique in two respects. Firstly, in the scale of the study and, secondly, in the attempt to 'close' the circle, that is to relate costs of nutrient reductions to associated benefits. Starting in the late 1960s, the empirical work on water-quality management has been focused on relatively small watersheds (Kneese and Bower, 1968). The linkage of economic activities with water-quality impacts has usually been done for a single sector, for example, waste water treatment plants, industrial effluents, or agricultural loads. During the last 20 years the main focus has been on coupling agricultural activities with pollutant loads (see Russell and Shogren, 1993, for an overview). In the project presented here, we consider all types of nutrient emission sources and try to trace their nutrient pathways in the Baltic Sea and in the surrounding nine countries. We found very few studies where the identification of management strategies are based on estimates of costs of pollutant loads to water bodies, ecological impacts, and their measurements in monetary terms. Admittedly, as will be evident from subsequent chapters, our aims cannot be achieved without several simplifying assumptions. One could therefore ask what a study like this has accomplished at all? Our answer is that decisions are actually taken at the international scale of the Baltic drainage basins and therefore analytical methods and techniques should be orientated to at least start to address the spatial and temporal scale problems involved in providing policy-relevant findings. The analysis presented in this book can therefore serve as an inventory of the type of information needed for the identification of efficient management strategies, and can provide relevant conclusions given the basis of available information.

The main focus of the book is on management strategies. Therefore, we will not discuss all the monitoring and, in particular, natural science research that has been carried out for a period of about 30 years (eg *Ambio*, 1990). Instead, we will give a brief background to the anthropogenic causes of degradation. Then, we will discuss the general requirements for the economic evaluation of management strategies. The remaining chapters will look at specific relevant topics.

The overall organization of the material has followed a 'pressure-state-impacts-response', (P-S-I-R) framework (Turner, Lorenzoni, et al, 1998; Olsen, Tobey and Hale, 1998). The P-S-I-R framework proved useful for the scoping of Baltic-related science, economics and management issues. It provided a way of identifying key issues, questions, data availability, land use patterns, proposed development and existing institutional frameworks, in a connected process (see Figure 1.2).

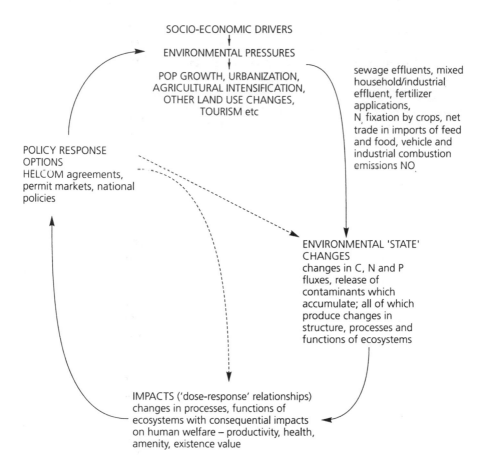

Figure 1.2 *P-S-I-R Cycle, Continuous Feedback Process*

The achievement of the aims of the book required expertise from the disciplines of, at least, ecology, economics and geography. When writing the book, all of the authors have attempted to avoid the jargon usually applied within their respective discipline. It is therefore our hope that the content of the book is generally accessible.

CAUSES OF BALTIC SEA ECOLOGICAL DEGRADATION

The state of the Baltic Sea is conditioned by many pressure variables, some of them shaped by natural processes and others influenced by economic activities. Eutrophication, – the effect of increased organic production due to the excessive supply of nutrients such as phosphorus and nitrogen (originating mainly from inland activities) – has been identified as the principal cause of the environmental degradation of the Baltic Sea and its coast (Sanden and Rahm, 1991; Wulff and Niemy, 1992). Since the turn of the century, the nitrogen and phosphorus contributed nutrient loads entering the sea have increased by about four and eight times respectively (Larsen et al, 1985). The most dramatic increases occurred after the second world war due to the introduction of artificial fertilizers (Nord, 1993, p12). The discharges increased until the 1980s when they stabilized at a fairly high level (Stålnacke, 1996). They decreased during the 1990s, although at a pace which, if sustained, will bring visible improvements only over the very long run. The Sea has also been contaminated by toxic substances including pesticide residues and heavy-metals transported from sources in the drainage basin. International bans on the use and emissions of many of these toxic substances were imposed in the 1970s. This has now resulted in a gradual recovery of eagle and seal populations (HELCOM, 1996). The effects of toxic substances on the Baltic ecosystem have been severe, however, and they still represent a serious threat.

Research, in terms of identifying and monitoring concentrations and the effects of toxic substances has contributed to such successful reduction measures. However, from a management point of view, the decisions are 'simple' when toxic substances accumulate that cause a reduction in human welfare. They should be banned or phased out. For nutrients causing eutrophication, the management decision is not so simple, as nutrients are not artificial substances, but important and essential components of the ecosystem. What is problematic is too great a nutrient loading spatially or temporally speaking.

The principal nutrient pollution sources for the Baltic Sea are municipalities, industries and agriculture, located both in the coastal zone and also beyond in the drainage basin. A range of pollution pathways can also be identified, which differ in terms of their linkages to the emission source and the associated impact on the Baltic Sea. They include diffuse 'non-point' sources such as airborne emissions and agricultural runoff; and 'point' sources from urban areas and industry. Thus, untreated sewage from some 30 million people combined with inadequately treated industrial effluent is still directly discharged into the sea or carried into it via rivers.

Inadequate or absent municipal sewage treatment in the eastern, southern and south-western sections of the drainage area pose problems that are exacerbated by the synergistic effects of untreated industrial effluent waste streams passing through the same facilities and into the rivers and the Sea. Agricultural practices, including intensive livestock husbandry, are a major contributor to the high nutrient load and consequent eutrophication problem. Current loads of nitrogen and phosphorus entering the Baltic Sea are at least three times those of the 1950s. The resulting excessive production of organic matter causes oxygen depletion when it decays and therefore threatens marine life. Eutrophication is now pronounced in the entire Baltic Proper basin, south of Åland Sea, particularly in the Gulf of Finland, Gulf of Riga and many coastal areas in the eastern, southern and south-western Baltic Sea areas. Massive blooms of plankton algae proliferate. The most conspicuous ones are those composed of cyanobacteria that cover large parts of the Baltic Proper in late summer. They are sometimes toxic and when they drift ashore create banks of foul-smelling detritus that inhibits bathing and recreational use of the beaches and sea. Drastic changes have occurred in many areas of the food chain. Fish populations have been depleted, partly due to fishing activity, partly due to eutrophication. For instance, the Baltic cod population has been on the verge of extinction due to intensive fishing, but also due to eutrophication that has caused oxygen depletion in those areas where these species can successfully reproduce.

The principal anthropogenic causes of the current degraded ecological functioning of the Baltic Sea can be classified into two categories: market failures and government intervention failures. Market failures occur when all resource inputs – labour, capital, and nature – for the production of goods and services are not subjected to market transactions. The most widespread type of market failure is that of pollution externalities, where, for example, nature's waste assimilative capacity is used free of charge in the production of marketed goods. In the case of the Baltic Sea, significant external costs result from waste generators (municipalities, industry and farms) who over-utilize the waste assimilative capacity of the environment, for example, rivers and the Baltic Sea itself. This environmental function is perceived to be virtually free of charge because of the absence of market prices. Some waste generators have also had almost open access to the marine waste repository.

Government intervention failures have also been partly responsible for the environmental degradation process in the Baltic. The effectiveness of the regulation of sewage treatment facilities and practices, for example, varies dramatically from country to country around the Baltic. There is a general absence of properly integrated coastal resource management policies and water catchment management and planning. This has resulted in intersectoral policy inconsistencies and resource depletion and degradation, with the loss of wetland ecosystems being one of the most significant outcomes. Coastal wetlands were ditched and drained to meet the demands of an expanding agriculture, and dredged or filled in to make room for urban and industrial development, including port facilities (Turner et al, 1999).

A related regulatory failure concerns the international coordination of Baltic Sea policies. The biophysical and socio-economic characteristics of the Baltic Sea and its drainage basin are such that a change in environmental pressure in one country can result in impacts across the entire basin. However, the target sources and areas for pollution abatement measures are much more concentrated, largely lying within the national boundaries of the 'transition economies' in the SEBR. Localised abatement action in one sub-basin can in some circumstances can also lead to increased damaged elsewhere in another sub-basin. This is the case with the Gulf of Riga, which is phosphorus limited, implying that increases in phosphorous loads create further damage in the bay while changes in nitrogen loads have insignificant impacts. Most parts of the Baltic Sea are nitrogen limited, thus implying that reductions in nitrogen loads improve the conditions of the Sea. Reductions in the phosphorous load of the Gulf of Riga in order to improve the bay water quality, implies the release and transport of nitrogen to the Baltic Proper, which is nitrogen limited. (Turner et al, 1999).

MANAGEMENT AND POLICY EVALUATION

The water quality decline in the Baltic Sea and the importance of the interdependence between the Baltic Sea countries in this degradation process were formally realized as early as the 1970s. The first international Convention on the Protection of the Marine Environment of the Baltic Sea Area was signed in 1974 by all the nations with Baltic coasts. Final ratification by all parties was achieved in 1980. The original signatories included seven countries: Denmark, Finland, West Germany, the German Democratic Republic, Poland, the Soviet Union and Sweden. After the collapse of the Soviet Union, Estonia and Lithuania joined the Convention in 1992, as did Latvia in 1994. This Helsinki Convention covered the entire Sea area out to the border between Kattegat and Skagerack. The objective was to protect the marine environment from general pollution, both land-based and from shipping. The Convention prohibits waste dumping at sea, as well as pollution from seabed utilization and research. Between 1990 and 1992 a more comprehensive joint environmental action programme was worked out at a political level. At the end of this process a new, more stringent, Convention was signed by Denmark, Estonia, Finland, Germany, Latvia, Lithuania, Poland, Russia, Sweden and the European Union (EU). The new Convention includes coastal and inland waters and defines the scope of the required ameliorative actions to be the whole Baltic drainage basin region.

The first decade of the Helsinki Convention did not prove to be very successful. The pollution of the Sea continued, and countries were not ready to cooperate effectively. The second decade saw several steps making the Helsinki Commission (HELCOM) – the executive body of the Convention – a more visible and increasingly important player in the game. Nonetheless, until the end of the 1980s, most of the activities were confined to monitoring and data collection. While this has not improved the quality of the Sea, it at least established fairly firm foundations for any subsequent actions. However, during the 1980s HELCOM transformed itself into a more action-oriented body (Hjort, 1992). This manifested itself in the early 1990s by the identification of 132 'hot spots' in the drainage basin that should be the focus for abatement investment (HELCOM 1993, No 46). However, in spite of many years of monitoring and the suggestions in the early 1990s to reduce the pollutant loads from the hot spots the Baltic Sea is far from being a 'healthy' sea. There are at least two important interrelated reasons for this failure. One is the lack of perspective and action over the interdependency between countries and the economic impacts of measures reducing the nutrient loads. The second is the absence of institutions and mechanisms for undertaking nutrient reduction investments. The ultimate purpose of this book is to help fill this knowledge gap by presenting results from a large-scale economic evaluation of policies for reducing nutrient loads to the Baltic Sea.

A typical evaluation method is economic cost-benefit analysis, which may be deployed as a conventional economic efficiency analysis, or in a constrained form as a cost-effectiveness analysis. In the standard cost-benefit method the traditional decision criterion is to maximize net economic benefits. Given that Baltic resources are being depleted or lost, the question to ask is whether the people living in the region should be concerned. The answer is yes, because many of the resources and their encompassing systems are very valuable assets, some of which can be quantified in monetary terms. The Baltic coasts are popular tourist areas, and also serve as spawning, nursery and feeding grounds for several species of marine and freshwater fish. The value of fish catches is some 540 million ECU per annum. This is, however, small in comparison with the recreation and amenity benefits that the Baltic's coast and other sites provide, or with the implicit value of the marine waste assimilation service (ie very large replacement costs).

Correctly applied the cost-benefit analysis will give two types of results. One is whether or not, on economic efficiency grounds, a pollution reduction project should be carried out, ie when benefits are at least equal to costs. The other gives information on the 'best' or most efficient allocation of different pollution reduction measures. This means a configuration of measures that generates a given reduction in the pollutant load to the Baltic Sea at minimum cost. The cost-benefit criterion is too narrow, however, in situations where not all resource values can be translated into monetary terms and when criteria other than economic efficiency are deemed important by the relevant decision makers. Operational trade-off relationships can nevertheless be developed by imposing constraints (for example, ambient environmental quality standards, regional employment/income targets, conservation of designated nature reserves) (Bower and Turner, 1998). This latter method is a cost-effectiveness analysis. In this book we will present results from both cost-benefit and cost-effectiveness analyses. In addition, we will investigate the economic impacts of different coordinated and uncoordinated policies for nutrient reductions by the Baltic Sea countries.

A cost-benefit analysis requires that the costs of nutrient reductions must be related to the associated benefits as measured in monetary terms. Although seemingly simple in principle, in practice this requires an enormous amount of information on the Baltic Sea and its drainage basin that can be classified into three categories:

1   *Pressure analysis* – process data linking emission sources with nutrient loads to the Baltic Sea. This includes firstly, the investigation of all emission sources and the quantification of their emissions, which depend on the characteristics of the land where the source is located. Next, the share of total emissions that reaches the Baltic Sea must be quantified for all sources.
2   *Impact analysis* – impacts on the Baltic Sea from changing nutrient loads and their evaluation in monetary terms. This implies the calculation of costs for nutrient reductions for all possible options, and an estimation of the benefits from associated reductions.
3   *Drainage basin wide policy response analysis* – if sufficient information is obtained under 1 and 2 (if a cost-benefit analysis can be carried out) this constitutes the basis for the simulation of different policies for improving the Baltic Sea. One important question to raise is therefore under what conditions all involved countries would agree on a common

nutrient reduction policy, and whether or not this is the best policy for all participating countries.

The ultimate aim of these three tasks is to link the economic activities that bear the costs of nutrient reductions with associated human welfare changes from impacts on the Baltic Sea. The provision of analysis sufficient to provide all the chain linkages is rare at such a large scale as in the Baltic Sea and its drainage basin, making the approaches and results presented in this book quite novel in this respect. Projects with similar aims have been applied only to small watersheds and not on a transnational basis. There are good reasons for this lack of similar large-scale studies. The information requirements for carrying out such a cost-benefit analysis are enormous, and several simplifying assumptions have to be made before the main analytical steps can be coupled together.

STRUCTURE OF THE BOOK

This book is organized according to the three main classes of information requirements in the P-S-I-R approach, and the content is therefore divided into three main sections: Part I Land Use and Nutrient Loads; Part II Economic and Ecological Evaluation of reductions in nutrient loads to the Baltic Sea; and Part III Institutions and Policies for Baltic Sea management. Part I contains two chapters where land use in the drainage basin is investigated and, based on this, nutrient loads are calculated. According to the estimates derived via a geographical information system (GIS) presented in Chapter 2, the total area of the Baltic Drainage Basin is approximately 1,745,000 km^2 and contains about 85 million people. The area displays a varied geographical nature. The north-west quadrant is largely mountainous terrain, the north-east quadrant is dominated by forests, wetlands and lakes, while the south and west quadrants possess significant agricultural areas and population concentrations. The total population of 85 million is unevenly distributed with more than 50 million on the southern side of the Baltic Sea. Other large population centres, such as the St Petersburg region (8 million inhabitants), are more spatially concentrated.

The particular combination of biophysical and socio-economic characteristics has important implications for the environmental vulnerability of the Baltic Sea and its resource system. We would expect the nutrient loads to be higher in the southern, densely populated regions than in the north. This is also verified by the load estimates that are presented in Chapter 2. These load estimates are unique in the sense that they relate coastal loads of nitrogen and phosphorus to different economic sectors – agriculture, traffic, industry and households – in each country. Admittedly, not all the necessary information on nitrogen transports for linking emission sources with loads in the coastal water is available in scientifically reliable forms. The coupling of nutrient loads to emission sources is therefore based on many simplifying assumptions concerning leaching and nutrient-retention processes. Nevertheless, the expectations of relatively high nutrient loads from the SEBR are confirmed by the load calculations presented in Chapter 2. Poland

and Germany together account for about 40 per cent of the total nitrogen load and approximately 66 per cent of the total phosphorous load. There are a number of differences between the nitrogen and phosphorus load. Agriculture is the largest source of nitrogen load, and accounts for approximately 45 per cent of the total load. Phosphorous effluents from sewage treatment plants make up 60 per cent of the total phosphorous load. Another important difference between the nitrogen and phosphorous loads is that a much larger proportion of the phosphorous load sources are located on or close to the coast. Furthermore, the total nitrogen load is about 25 times higher than the phosphorus load. These differences imply that the impact of a given nutrient load reduction in the Baltic Sea is larger for phosphorus than for nitrogen, which therefore implies lower costs for nutrient load reductions.

The nutrient load would probably be higher if it were not for the remaining wetlands, however, in particular in the southern part of the drainage basin. In Chapter 3, the nutrient sink capacity of the Baltic Sea wetlands is calculated. The estimates are based on GIS data and statistical sources. The GIS data base shows that about 138,000 km² of the basin is covered by wetlands, which corresponds to 8 per cent of the total drainage basin area. It is estimated that these wetlands contribute to the nutrient reduction target by creating nitrogen sinks corresponding to about 15 per cent of the total nitrogen load. Thus, without the nitrogen sink functioning of these wetlands, costly measures would have to be implemented to gain the same nitrogen reduction. It has also been calculated that a further nitrogen reduction of about 10 per cent can be achieved if drained wetlands were to be restored.

In Part II, a cost-benefit analysis is carried out. It contains three chapters: calculations of the costs of reduction in nutrient loads, are first presented, then a marine ecological simulation of the impacts of changes in nutrient loads in different Baltic Sea basins is illustrated; and finally, the valuation of these changes in monetary terms is explained. The calculation of costs in Chapter 4 by Gren is to a large degree based on the load estimates presented in Chapter 2, which is combined with the estimates of all possible measures for reducing the nutrient loads to the Baltic Sea, and the calculation of their impacts and costs. In principle, three classes of nutrient-load reduction options are included:

1 process-related or end-of-pipe measures which reduce the emissions at source;
2 land use measures which decrease the leaching of nutrient, for a given level of emissions; and
3 retention-oriented options, such as buffer strips and wetlands that will reduce the load to the Baltic Sea from upstream emissions and leaching.

The calculation of minimum costs for about 15 different measures for reducing nitrogen loads in each country, shows that a combination of these classes of measures yields the least cost solutions for given nitrogen reduction targets. The improved functioning of sewage treatment plants located on the coast, restoration of wetlands and arable land use change measures are the least costly options. Another important result is the rapid increase in cost at levels above 40 per cent reduction targets. At these high reduction targets, only the most expensive options are available, and the minimum costs increase from about SEK 15 million per year to SEK 30 million when the nitrogen reduction target is increased from 40 to 50 per cent. The calculations also show that the minimum costs of phosphorous reductions correspond to about 20 per cent of the cost of the same nitrogen reductions as measured in percentage decreases. A third factor influencing the lowest cost solutions is whether or not simultaneous or separate reductions in nitrogen and phosphorus are carried out. The reason for differences in costs in this regard is that the class 2 and 3 abatement measures mentioned above have impacts on both nutrients. Under simultaneous reduction policies this feature implies a cost advantage compared to other measures targeted at a single nutrient. The total cost of a 50 per cent nutrient abatement policy is reduced by about 5 per cent if simultaneous reductions in nitrogen and phosphorous are carried out.

In order to complete a cost-benefit analysis, we also need to compare the costs of nutrient reductions with the benefits obtained from these reductions. This means that we must first relate different nutrient reductions to the biological conditions in the various basins of the Baltic Sea. Then, these impacts have to be 'translated' into terms meaningful to people so that they are able to assign monetary values to these changes. An attempt to relate nutrient loads to their impact in the Baltic Sea is carried out in Chapter 5 by Wulff. This is broken down into two steps, the estimation of the relation between nutrient loads and nutrient concentration ratios in different basins, and the relation between nutrient concentration ratios and the biological functioning of the basins. The available data on the second step is somewhat qualitative. Thus, it is known that alterations in the nitrogen/phosphorus ratio increase the production of certain algae. This can further generate changes in the composition of fish species and also create sea-bottom conditions devoid of any biological life.

This qualitative information is not enough for an appropriate valuation, which requires data on *how much* fish and other species change due to alterations in the nutrient loads. Therefore, we have to rely on a simulation of impacts related to nitrogen and phosphorus concentration ratios to estimate different percentage reductions in the nutrient loads. An important result from the simulation exercise is that the stabilization of new concentration ratios requires about 5–10 years for nitrogen reductions and about 15–25 years for phosphorous reductions. This means that we must be patient when expecting results from nutrient reduction programmes. Another important result concerns the spatial relationship between water basins. A reduction in the nutrient load of the Baltic Proper and the Bothnian Sea will have impacts on the concentration ratios, not only there, but also in the entire Baltic Sea. These relationships between basins, and also between countries with coasts in the different basins, provide one important aspect of the public good character of the Baltic Sea. The publicness of the Baltic 'resource' constitutes a difficult challenge for the valuation and sustainable management of the Sea.

The long-term nature and the spatial complications of nutrient load reduction make it difficult to carry out a study where reductions are valued in monetary terms. Some commentators would, in any case, argue that the Baltic Sea has intrinsic environmental value and that it also has value because it is a shared heritage to be conserved and managed as a common resource for current and future generations. These motivations are related to other types of environmental value, beyond the use values that underlie the waste assimilation, recreation and fisheries functions and outputs. Economists conceive of a concept called 'total economic value' which encompasses both use values and so-called non-use (existence and bequest) values, but excludes intrinsic value in nature, independent of human welfare. (Turner and Pearce, 1993; Gren ct al, 1994).

The standard economic approach to environmental-resource valuation is based on an individual, human 'preference-based' value system, in which the benefits of environmental gain (or the damages of environmental loss) are measured by total economic value, as discussed above. While debate continues over the precise components of total economic value and their individual boundaries, the economic approach to environmental-resource valuation is based on some agreed assumptions. It is assumed that individuals (organized as households) maximize utility (satisfaction). Given this, willingness-to-pay (ie their assigned valuation) for resources is a function of prices, income, tastes, attitudes household size, education and so on. The social value of environmental resources committed to some use is simply then the aggregation of private values.

In Chapter 6, such an approach, the so-called direct valuation method, is applied for assessing the willingness to pay for an improvement of the Baltic Sea to bring it back to the biological conditions prevailing prior to the 1950s. Studies are undertaken in two countries: Sweden and Poland. A random sample is chosen in each country and the respondents are asked if they are willing to pay a certain level of a tax each year in order to regain the Baltic Sea conditions of the 1950s. In this scenario it is further suggested that cost-effective programmes will be implemented to achieve this target. The results reveal a willingness to pay around SEK 3000 per person per year for Swedish citizens and SEK 300 for Polish citizens. When transferring these calculations to other Baltic countries, the total annual value amounts to about SEK 31,000 million. It is interesting to note that this is roughly the same amount as the cost of reducing the total anthropogenic and non-anthropogenic load of nitrogen in the Baltic Sea by 50 per cent.

Part III contains three chapters discussing policy and institutional arrangements for implementing nutrient-reduction targets. The first chapter by Markowska, Zylicz and Gren, discusses the design of a mechanism for the cost sharing for a programme to reduce the load of nitrogen by 50 per cent. They apply the so-called Chander-Tulkens model of sharing the cost of the provision of an international public good, such as the environmental preservation of the Baltic Sea. This model suggests that the costs should be shared among countries in proportion to their benefits from the provision of the good. The authors therefore use the benefits and cost estimates from previous chapters and demonstrate the outcome of such a cost-sharing mechanism. Predictions of cost-sharing schemes obtained from the Chander-Tulkens model are tested with actual data on abatement expenditures and international assistance. Several recommendations on how to optimize the Baltic-wide clean-up programmes are then made.

Chapter 8 by Gren, investigates whether the 50 per cent reduction target is the 'best' reduction target for the participating countries. The 'best' target is defined as the reduction level which maximizes overall net benefits. The extent of total net benefits is, however, determined by how countries coordinate their policies. Net benefits are therefore calculated and compared for coordinated and uncoordinated policies between the countries. Under the coordinated scenario, the total net benefits of all the countries are maximized, while in the uncoordinated case each country maximizes its own net benefits. Whether or not these policies imply large differences in net benefits between countries, depends on their nitrogen-reduction costs, and the marine transport of nutrients in the Baltic Sea. According to the results, the countries could together receive a maximum gain of about SEK 12 billion per year by switching from an uncoordinated to a coordinated policy. Total nitrogen loads then decrease by 40 per cent, as compared to approximately 20 per cent when each country acts on its own. The main reason for this difference is the inclusion of Polish environmental benefits that are mainly obtained by other countries due to Polish open coasts which transport most of the nutrient reductions beyond Poland's borders. Thus, under an uncoordinated nitrogen reduction regime, these countries will reduce much less than is optimal for the whole region. The richer countries may therefore have to provide the international public good themselves, letting the poorer countries 'free ride' to a greater or lesser extent (Sandler, 1998).

It is also interesting to note that the optimal total nitrogen reduction amounts to about 40 per cent, which is the level where the cost of nitrogen reductions start to increase relatively rapidly. This also means that at nitrogen reductions exceeding 40 per cent, costs increase sharply, so that at the level of 50 per cent nitrogen reduction, which is the ministerial target, there is a net loss of about SEK 15 billion.

In Chapter 9, Johannesson and Randås discuss the option of permit markets as a mechanism for allocating nitrogen reduction targets among four different regions, Denmark, Poland, the Baltic States and Sweden. The novelty of this work is in the consideration of the indirect effects on the general economies from emission reductions in nitrogen intensive sectors such as agriculture and transport. Computable general equilibrium (CGE) models are used to estimate the adjustments that take place in the economies under two classes of permit markets – national markets and a common regional market. The results indicate that the costs will be high under all systems, but that they are mitigated by a common regional market. Not all individual countries will gain from this form of cooperation. As in the simulation game involving all Baltic countries in Chapter 8, Poland, for example, is a net loser as compared to a national permit market. However, the results also show that only small changes from the 'grandfathering principle' (reduction requirements being proportional to initial nitrogen loads) are required to make all countries better off with a common permit market than national markets.

PART I

LAND USE AND NUTRIENT LOADS

2   LAND USE, POPULATION AND NUTRIENT LOADS

Ing-Marie Gren

The calculation of the costs of different nutrient-reductions strategies for the Baltic Sea requires information on the human activities generating the nitrogen and phosphorous emissions and their locations. Currently, data sets exist on nitrogen and phosphorous loads measured at about 90 different river mouths (for example, Stålnacke, 1996). Data on fertilizer use, arable land use, nitrogen oxides and ammonia emissions is also available for different countries in the drainage basin (Elofsson, 1997; Sandnes, 1993). However, the country and county borders of the nutrient emission generating activities do not coincide with the boundaries of the drainage basins. Therefore, the nutrient emission data are not closely correlated with the loads entering the Baltic Sea. Before any calculations of costs and simulation of policies can be carried out, such linkages must be made. In addition to national statistics, geographical information on the allocation of land use and population is a powerful tool to enable this linkage, but must be used in conjunction with other information. Data on the chemistry and biology of nutrient transformation are also necessary. The approach used to link nutrient loads with emission sources can be divided into two steps:

1   a geographical information system (GIS) description of land use and population patterns that enables the analyst to identify sources; and
2   augmentation of the GIS is by data from various official statistical sources on input use, such as fertilizers and petroleum, that generate nitrogen and phosphorus emissions.

LAND USE AND POPULATION IN THE DRAINAGE BASIN

Sweitzer et al (1996) present the results from the development of a GIS for the Baltic basin. This database is also available on the Internet http://www.grida.no/baltic. In total, the drainage basin covers an area of 1,745,000 km², spread across 14 different countries. The constructed map of the drainage basin is presented in Figure 2.1.

Figure 2.1 *The Baltic Sea Drainage Basin*

From Figure 2.1 we can see that Poland, Sweden and Finland cover a large share of the entire drainage basin and six other countries have small parts of their territories within the drainage basin (large part... Belarus, Czech Republic, Germany, Norway, Slovakia and Ukraine). The Swedish and Finnish regions also cover a large share of the total forest area of the drainage basin (see Table 2.1).

Almost 50 per cent of the area of the entire basin is covered by forest, most of which is located in the drainage basins of Bothnian Bay and the

Table 2.1 *Land Cover in the Baltic Sea Drainage Basin, Thousand km²*

Region	Forest	Arable land	Pasture	Inland water	Populated area	Other land	Total	Total in %
Bothnian Bay	193.6	7.1	0.3	11.7	0.6	55.2	268.6	15
Bothnian Sea	151.2	11.7	0.7	12.2	1.3	62.5	227.4	13
Gulf of Finland	223.7	36.0	14.9	56.2	1.9	83.4	416.1	24
Gulf of Riga	51.2	37.9	16.2	2.6	0.9	29.3	135.7	8
Baltic Proper	164.3	224.7	66.6	13.1	6.2	111.9	573.9	33
Danish Straits	1.6	16.5	3.3	0.5	1.1	7.3	29.7	2
Kattegat	50.1	19.1	2.0	10.5	1.1	11.0	93.7	5
Total	*835.9*	*352.9*	*104.0*	*106.9*	*13.1*	*332.2*	*1745*	*100*
in %	*48*	*20*	*6*	*6*	*1*	*19*	*100*	

Bothnian Sea. The second largest use is arable land, which accounts for 20 per cent of the total area and which is almost entirely located in the drainage basin of the Baltic Proper. Since agriculture is an important nutrient emission source, we would expect that changes in agricultural practices should be important options in any efficient nutrient reduction strategy. Another important source of nutrient effluent is households and industry. However, we see from Table 2.1 that only 1 per cent of the Baltic drainage basin region is densely populated (at least 200 inhabitants per km²). The entire region holds about 85 million people, most of which live in small towns/cities in the drainage basin of the Baltic Proper, which can be seen from Figure 2.2.

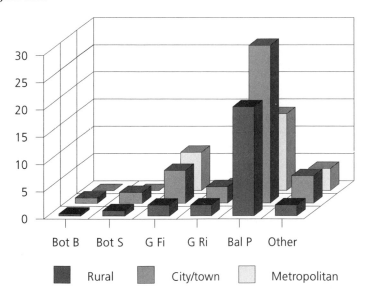

Key: Bot B = Bothnian Bay; Bot S = Bothnian Sea; G Fi = Gulf of Finland; G Ri = Gulf of Riga; Bal P = Baltic Proper; Other includes Danish Straits and Kattegat

Figure 2.2 *Urban and Rural Population in the Drainage Basin*

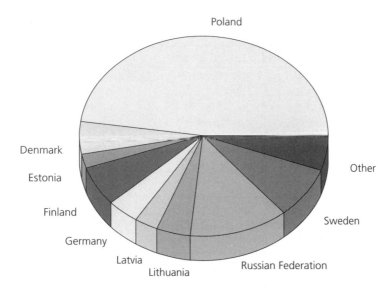

Figure 2.3 *Allocation of Population in the Drainage Basin Countries*

A metropolitan area is defined as a settlement with a population greater than 250,000. The population in a small city/town is between 200 and 250,000, while rural areas contain settlements with populations of less than 200.

The categorisation of people into urban and rural living space has implications for the nutrient-reduction measures that can be deployed. For densely populated areas, there are well-developed technologies for the construction of sewage treatment plants, which, as will be discussed in Chapter 4, provide a low-cost measure for reducing both phosphorous and nitrogen loads in the Baltic Sea. It is more difficult and costly to abate emissions from scattered populations unconnected by pipelines to large sewage treatment plants. Currently, about 70 per cent of the Basin's population lives in metropolitan areas or in small cities and towns.

When classifying the population according to nationality we find that almost 50 per cent live in Poland (Figure 2.3), while Poland, Sweden and Finland together account for 75 per cent of the total population. In the absence of advanced effluent-treatment capabilities for effluent generated by households and industry in their areas, we would expect a relatively high load of nutrients from these sectors.

Four countries – Poland, the Russian Federation, Finland and Sweden – together account for 80 per cent of the total area of the entire drainage basin (Figure 2.4).

NUTRIENT LOADS TO THE BALTIC SEA

Figure 2.1 shows that there are five countries with only small parts of their domestic territory located within the drainage basins, namely Norway, large part... Belarus, the Czech Republic, Slovakia and Ukraine. The total

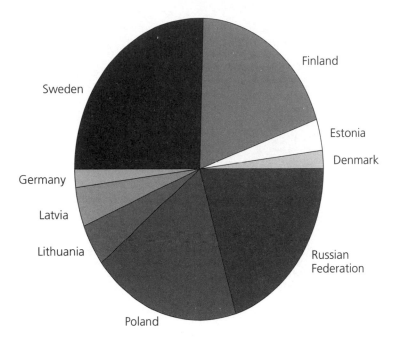

Figure 2.4 *Country Allocation of the Drainage Basins' Land Area*

area of these regions corresponds to nine per cent of the entire drainage basin. When relating nutrient loads to emission sources these regions are excluded because of the relatively high level of analytical and data calculations that would otherwise be required. Another reason for their exclusion is that these regions do not have any coastal zones adjacent to the Baltic Sea and their interest in the cleaning up of the Sea might therefore be small. The nine remaining countries account for 91 per cent of the area of the drainage basin.

The first step in the process of linking emission sources to nutrient loads in these nine countries is to determine a sub-division which satisfies the availability of natural science and economics data. We would, for example, expect nutrient transformation to be quite different in the south of Sweden than in the northern areas where there is a colder climate. Further, when simulating the impacts of different policies in reducing the load of nutrients, we also require the regional divisions that coincide with the relevant administrative boundaries. On the basis of these considerations, the drainage basin has been partitioned into 14 regions. All the regions coincide with country borders and with the three main basins of the Baltic Sea (Figure 2.5).

We are now equipped with a comprehensive description of the land use and population of the different countries and drainage basins. From this information it can be seen that nutrient loads from agriculture, households and industry are relatively high in countries located in the drainage basin connected to the Baltic Proper. In particular, since one-half of this sub-basin's population is located in Poland we should expect to have large loads relating to household and industrial activities. However, the load to the

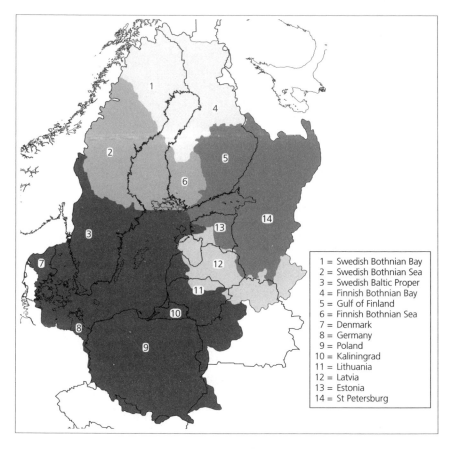

Figure 2.5 *Drainage Basins of the Baltic Sea*

Baltic Sea is determined, not only by the source emissions, but also by the characteristics determining the transformation of nutrients during transportation from the source to the coastal waters of the Sea. These, in turn, are determined by climatic, hydrological, biological and chemical factors in the wider landscape and the transporting water bodies within which the nutrients are deposited. The data available on these processes are neither reliable nor comprehensive.

Given the rate of deposition of nutrients on land the calculation of the load which finally reaches the Baltic Sea can be divided into two parts. First, we have to determine the form and rate of the leaching of nutrients deposited on land, which can be done via manure spreading, fertilizer applications and atmospheric deposition. The amount of leaching also depends on differences in land use. We would, for example, expect less leaching from land covered with perennial crops than from land that is covered by crops for only part of the year. Given a certain rate of leaching, the next part constitutes the surface and subsurface transport of the nutrient leaching. The share of the leaching nutrient that finally reaches the Baltic Sea will then depend on the location of the leaching in relation to the distance to the coast and downstream landscape composition and other climatic and hydrological characteristics. During the transport, nitrogen is transformed in three ways:

1 assimilation by plants;
2 sedimentation on stream bottoms or lakes; and
3 denitrification where nitrogen is transformed into nitrogen gas.

The transformation of phosphorous takes place by assimilation and sedimentation, therefore we would expect nitrogen retention – the leached nutrient that does not reach the Baltic Sea coast – to be higher, the greater the distance from the emission source to the coastal water.

Ideally, we would have a functional description of nutrient leaching and retention from all nutrient emission sources and their locations in the drainage basin. However, according to Elofsson (1997) such functions are not available on the Baltic Sea drainage basin scale. They are only calculated for small field and watershed conditions. Therefore, very simple, linear relations between nutrient deposition, leaching and retention have been assumed. Furthermore, no distinction could be made between nitrogen and phosphorous. For a comprehensive presentation we refer to Elofsson (1997). The resulting ratios between nutrient (nitrogen and phosphorous) deposition and leaching, leaching and retention, and finally, nutrient deposition and load to the Baltic Sea for the different regions are as presented in Table 2.2. According to the information in Table 2.2, Denmark and Sweden are the countries that produce the highest impact on the Baltic Sea from a unit, say one ton, nutrient emission reduction.

Table 2.2 *Ratios of Nutrient, Leaching, Retention and Load to the Baltic Sea*

Region	Leaching/ deposition	Retention/ leaching	Load/ deposition	Deposition to the sea/air emission	
				NO_x	NH_4
Finland:					
Bothnian Bay	0.16	0.3	0.11	0.03	0.03
Bothnian Sea	0.16	0.3	0.11	0.03	0.03
Gulf of Finland	0.15	0.3	0.10	0.03	0.03
Sweden:					
Bothnian Bay	0.07	0.25	0.04	0.13	0.13
Bothnian Sea	0.10	0.25	0.07	0.14	0.13
Baltic Proper	0.21	0.33	0.14	0.16	0.16
Russian Federation:					
Kaliningrad	0.27	0.8	0.05	0.03	0.02
St Petersburg	0.20	0.8	0.04	0.03	0.02
Poland	0.23	0.8	0.05	0.06	0.05
Lithuania	0.11	0.8	0.02	0.06	0.05
Latvia	0.15	0.8	0.03	0.09	0.08
Estonia	0.14	0.8	0.03	0.10	0.12
Denmark	0.22	0.3	0.10	0.15	0.14
Germany	0.19	0.3	0.06	0.06	0.04

Source: Elofsson 1997, p27, and Gren et al (1995)

In principle, these countries could achieve a maximum reduction of 10 or 14 per cent respectively. Table 2.2 also reveals that, in spite of relatively high initial nutrient leaching in Poland and the Russian Federation, the final impact on the Baltic Sea is relatively small for both these countries. The reason is the large fraction of leached nutrient, 80 per cent, that apparently never reaches the coastal water. Given that these retention rates are correct, nutrient reduction in these countries would mainly be associated with improved inland water quality and would therefore be of less significance for the Baltic Sea. However, it should be kept in mind that the numbers on retention rates are not based on appropriately scaled 'field' measurements and are therefore highly arbitrary. The importance of the level of retention rates will become obvious from the cost estimates in Chapter 4.

The last two columns in Table 2.2 provide information on the share of emissions of nitrogen oxides and ammonia respectively, originating in each region, that is deposited in the Baltic Sea. Excluding southern Sweden, these shares are relatively low. We have, however, more trust in these data than in the retention figures since they are based on 10 years monitoring of atmospheric transport (Sandnes, 1993).

Now that we have information on the shares of depositional nutrients which reach the Baltic Sea, it remains to estimate nutrient deposition on land and inland waters in the different countries in the drainage basin. There are three types of deposition pathways:

1   fertilizers and manure on arable land;
2   household and industry sewage entering inland waters; and
3   atmospheric deposition.

The atmospheric deposition of nutrients is linked to energy combustion and ammonia releases when manure is deposited on land. A number of data sources have been used for estimating the emissions and deposition from these sources. In the case of atmospheric deposition we also need to know how the nitrogen oxide and ammonia emissions from each country are distributed across all countries within the drainage basin. Here, we make use of transport matrices, which contain numbers for pollutant transports between different regions and countries (Sandnes, 1993). The total calculated anthropogenic loads of nutrients amount to 728,000 tons of nitrogen per year and approximately 37,000 tons of phosphorus per year. The inclusion of background leaching and air emissions from countries outside the drainage basin brings the calculated total nitrogen load to about 1,000,000 tons of nitrogen per year (Gren et al, 1995). Other published estimates of nutrient loads vary between 400,000 and 1,400,000 tons of nitrogen and 32,000 and 70,000 tons of phosphorus (Helcom, 1993; Stålnacke, 1996). These estimates are based on measurements of nutrient concentrations at different river mouths along the Baltic Sea coasts. However, there is much uncertainty in our estimates, which implies that, although well within the range of other estimates, they must be interpreted with caution.

Polish nutrient emissions are undoubtedly the most important source of both nutrient loads, and correspond to 33 per cent of the total nitrogen load and 50 per cent of the total phosphorus load. The relatively large load of nitrogen from Germany is explained by the inclusion of air emissions, not only from the drainage basin, but also from the rest of the country.

In Figures 2.6 and 2.7 we present the calculated country allocations of nitrogen and phosphorus loads respectively.

The four countries – Poland, Finland, Sweden and Germany – together account for about 75 per cent of the total loads of both nitrogen and phosphorous. However, their shares of the total loads are different. In particular, Polish and German shares differ between the two nutrient loads. A high relative share of German nitrogen loads, about 20 per cent, is explained by air borne nitrogen emission from the entire country, rather than from the area of the country that coincides with the borders of the drainage basin. The relative German share of total phosphorous loads is much lower than is the case for nitrogen, and instead, Poland accounts for one-half of the total phosphorous loads. This is explained by the large population and a low level of phosphorous removal at Polish sewage treatment plants. In spite of a long tradition of phosphorous stripping in sewage treatment plants, the loads from Sweden and Finland are still considerable. This is explained by the runoff from intensively-cultivated arable land as it is fertilized. The role of agriculture is more pronounced when considering nitrogen leaching from land and air transports of ammonia. The agricultural sector then accounts for 50 per cent of the total nitrogen load, which can be seen in Figure 2.8.

The agricultural sector accounts for slightly more than 50 per cent of the total load of nitrogen and 33 per cent of the total load of phosphorus. The agricultural load of nitrogen includes deposition due to fertilizer and manure

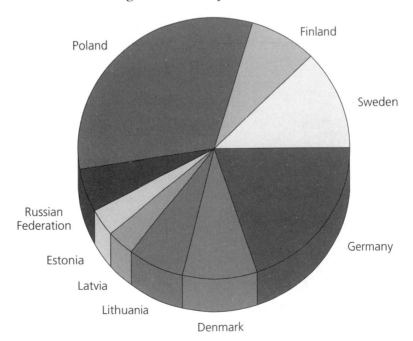

Figure 2.6 *Nitrogen Loads from Different Countries*

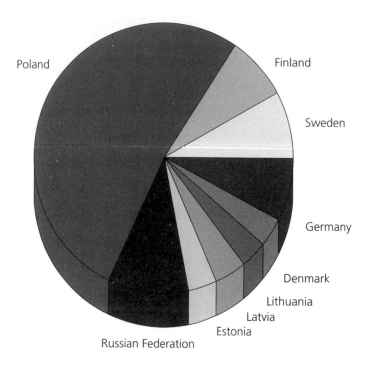

Figure 2.7 *Phosphorus Loads from Different Countries*

applications. The latter also generates ammonia emissions, directly on the Baltic Sea and via surface water flows across the drainage basin. The direct-deposition of ammonia into the Baltic Sea, however, constitutes the main part of the load from ammonia emissions, 66 per cent of the ammonia load.

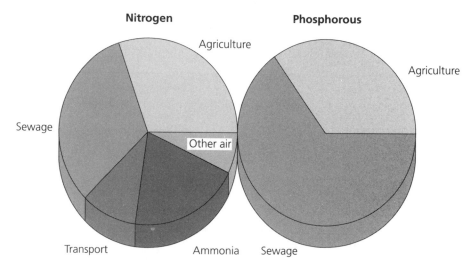

Figure 2.8 *Nitrogen and Phosphorus Loads from Different Sources*

Table 2.3 *Nitrogen and Phosphorous Anthropogenic Loads to the Baltic Sea*

Region	Nitrogen, thousand tons per year				Phosphorus, thousand tons per year		
	Fossil fuel	Agriculture	Sewage	Total	Agriculture	Sewage	Total
Sweden:							
Bothnian Bay	1.9	1.3	2.6	5.8	0	0.5	0.5
Bothnian Sea	6.3	4.5	3.1	13.9	0.2	1.0	1.2
Baltic Proper	11.5	31.4	27.7	70.6	0.5	0.6	1.1
Finland:							
Bothnian Bay	1.1	8.6	5.8	15.5	0.7	0.8	1.5
Bothnian Sea	1.6	11.4	8.6	21.6	0.9	0.4	1.3
Gulf of Finland	1.1	11.8	7.8	20.7	0.5	0.4	0.9
Russia:							
Kaliningrad	1.0	5.1	9.9	16.0	0.1	0.5	0.6
St Petersburg	9.6	7.7	4.5	21.8	0.3	2.5	2.8
Poland	21.8	108.5	102.9	233.2	6.5	11.9	18.4
Lithuania	2.4	19.7	20.6	42.7	0.4	0.7	1.1
Latvia	2.0	13.7	9.7	25.4	0.7	0.7	1.4
Estonia	2.3	9.8	12.0	24.1	0.1	1.1	1.2
Denmark	11.9	49.9	10.5	72.3	0.9	0.8	1.7
Germany	50.5	86.9	7.0	144.4	1.0	1.5	2.5
Total	125.0	370.3	232.7	728.0	13.1	23.7	36.8

The calculated air transports from the combustion of oil products and coal also include direct deposition and water transports which correspond to 17 per cent of total nitrogen load. Other air transports include the emission of nitrogen oxides from stationary combustion sources, which account for 7 per cent of the total nitrogen load. From Figure 2.8 we can also note that sewage from households and industry are important sources of both nutrient loads and account for 33 per cent and 66 per cent of the total load of nitrogen and phosphorus respectively. The overall allocation of nutrient loads in terms of countries and sources is presented in Table 2.3.

In the light of the information in Table 2.3 we would therefore expect that measures to reduce the nutrient loads from the agricultural sector, as well as the nutrient content in sewage from households and industry, will play an important role in any cost-effective nutrient-reduction programme. Further, the large share provided by Polish nutrient loads implies that the implementation of abatement measures in this country will be significant. However, whether or not this occurs also depends on the financial costs of different measures and their final impacts on the Baltic Sea.

3 WETLANDS AS NUTRIENT SINKS

Carl Folke and Åsa Jansson

The role of wetlands as nutrient sinks has received a lot of attention (*Ambio*, 1994; Etnier and Guterstam, 1991; Ewel, 1997; Hammer, 1989; Mitsch and Gosselink, 1993; Mitsch and Jörgensen, 1989; Nichols, 1983). The reason for analysing this functional capacity is the serious eutrophication of the Baltic Sea. In Sweden, large amounts of money are spent on sewage treatment plants to further increase nutrient-removal capacity. Sewage treatment plants cannot reduce non-point source pollution howeren. Using wetlands as filters between land and sea is an increasingly-applied abatement technology (for example, Baker 1992; Etnier and Guterstam 1991; Ewel and Odum 1984; Knight 1992, Mitsch 1992; *Ambio*, 1994), and appears to be cost-effective (Andréasson-Gren et al, 1991; Gren, 1995a, and Chapter 4 in this volume). Also, wetland restoration seems to provide a good way to remove nitrogen (Cooper and Findlater, 1990; Hammer, 1989; Leonardsson, 1994; Mander and Mauring, 1995).

There have been several site-specific studies of wetlands as nitrogen filters in the Baltic Sea drainage basin (Jacks et al, 1994; Jansson et al, 1994). There have also been studies of specific large wetlands in sub-drainage basins of the Baltic area, in Louisiana, and the Everglades (Costanza et al, 1990; Fitz et al, 1993; Krysanova et al, 1989). But, this study is, to the best of our knowledge, the first that investigates the nutrient removal capacity of a mosaic of wetlands in a large-scale drainage basin.

In this chapter we will present an analysis of the capacity of existing natural wetlands in the Baltic Sea drainage basin to serve as nutrient sinks, in particular, for nitrogen. The first part of the chapter presents a geographical information system-based (GIS) wetland distribution map of the whole region, with information about wetland distribution within countries as well as the loss of wetland areas that has occurred in recent decades. In the second part, we estimate the retention capacity of the atmospheric deposition of nitrogen by the wetlands in the Baltic Sea drainage basin. We also estimate their potential as filters of nitrogen arising from the human population, in terms of excretory release in processed sewage water.

There are still large gaps in understanding of the role of wetlands in the Baltic Sea drainage basin. Their links to other ecosystems such as agriculture and forests, to water and nutrient flows between systems and to human activities affecting these systems, have yet to be properly investigated. Due to lack of information on basic hydrology and topographic location, we could not estimate, for example, how much of the leakage from forests and agricultural ecosystems can be retained by wetlands. The nitrogen retention capacity that we estimate is compared to the annual inflow of nitrogen to the Baltic Sea. The third part of the chapter provides results from an 'ecological footprint' analysis of the area of wetlands that the largest cities within the drainage basin appropriate for processing human nitrogen emissions. We end the chapter with a summary of the major results. The chapter is based on Sweitzer et al (1996), Folke et al (1997) and Jansson et al (1998).

WETLAND DISTRIBUTION WITHIN THE BALTIC SEA DRAINAGE BASIN

The lack of information on the amount and distribution of wetlands in the Baltic drainage basin area is noteworthy in light of the interest and importance of the capacity of these ecosystems to filter and retain nutrients. A wetland map of the Baltic drainage basin was derived using a combination of statistical and spatial data sources based on the GIS database mentioned in Chapter 2. Wetland acreage was obtained from statistical sources on various administrative levels in the 14 countries of the drainage basin. Generally, wetland statistics were only available on more aggregated levels than agricultural statistics, which may reflect societies' perception of the relative economic importance of agricultural land versus wetlands. We relied on statistical data to provide estimates of wetland areas, and used other map layers from the database to estimate the spatial distribution of the wetland areas. Although the derived wetland map has a coarser resolution than the other derived map layers it is, to our knowledge, the best available digital data set of the distribution of natural wetlands in the Baltic Sea drainage basin.

Table 3.1 *Present and Drained Wetland Area by Country in the Baltic Sea Drainage Basin*

Country	Present wetland area (km²)	Drained wetland area (km²)
Belarus	3723	
Denmark	554	1034
Estonia	10,092	3364
Finland	42,092	98,215
Germany	576	NA
Latvia	3161	18
Lithuania	1154	NA
Norway	2206	NA
Poland	13,124	118,116
Russian Federation	11,700	NA
Sweden	48,941	1235
Total	*137,723*	*221,982*

Source: cited in Jansson et al (1996)

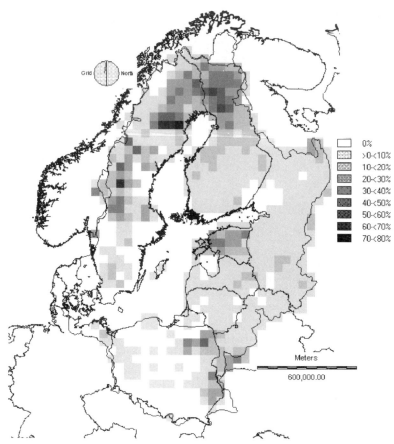

Source: Map derived from the GIS database created by Sweitzer et al (1996)
Note: Each square represents an area of 50x50 km.

Figure 3.1 *Distribution of Wetlands in the Baltic Sea Drainage Basin*

The results show that there are approximately 138,000 km² of natural wetlands in the Baltic Sea drainage basin (Table 3.1).

Wetlands cover roughly 8 per cent of the drainage basin and are most prominent in the northern regions. The majority of wetlands are found in the Bothnian Bay drainage basin (41 per cent). The Bothnian Sea, Gulf of Finland and Baltic Proper drainage basins each have between 16 and 17 per cent of the total wetlands, while the remaining drainage basins contain fewer than 6 per cent of the wetlands (Sweitzer et al, 1996). Wetland ecosystems have been seriously effected by human activities. During the last century vast areas of wetlands have been drained, and in some countries as much as 90 per cent of all wetlands have been lost (Rydlöv et al, 1991). The distribution of wetlands by country is given in Table 3.1, together with data on drained wetlands for those countries where such statistics were available. The geographical distribution of natural wetlands is presented in Figure 3.1.

ESTIMATING NITROGEN RETENTION BY WETLANDS IN THE BALTIC SEA DRAINAGE BASIN

Wetlands are multifunctional in the sense that they generate several eco-system services such as supplying habitat for many plants and animals (including endangered species), mitigating floods, recharging aquifers, and improving water quality by removing organic and inorganic nutrients and toxic metals from the water that flows across the wetlands (Mitsch and Gosselink, 1993). Here we focus on the annual nitrogen-retention capacity of wetlands in the Baltic Sea drainage basin. By retention, we mean the annual uptake of nitrogen by a wetland.

Using the GIS data base of the Baltic drainage basin, we estimate the amount of nitrogen retained by natural wetlands in three scenarios:

1 Wetland retention taking only atmospheric deposition of nitrogen into account. Atmospheric nitrogen deposition includes both wet and dry deposition, and indirectly consists of nitrogen emissions from human activities such as transportation and livestock farming. Data on total nitrogen deposition were derived from Iversen et al (1991). The data are mean measurements for 1985, 1987, 1989 and 1990. To be comparable with the wetland distribution map, the data were transferred from the 150x150 km EMEP grids into 50x50 km grids. This scenario provides a minimum retention estimate.

2 Calculating potential retention capacity by adding to atmospheric deposition, the nitrogen in excretory release from the human population. Concentration of nitrogen in urine and faeces from the human population is approximately 4 kg nitrogen person $^{-1}$ y $^{-1}$ (Torell, 1997). We assume that excretory release is processed in sewage treatment plants, and that wetlands receive the nitrogen in water discharged from these plants.

   Alternative techniques, such as septic tanks, also remove anthropogenic nitrogen discharge. These alternatives are not included in the study. At present, several cities are, or will be, connected to sewage treatment plants. Several municipalities have created wetlands to process sewage water (for example, Klochak, 1993). Based on empirical data (Gren et al, in press) we assume that the nitrogen in excretory release was reduced by 20 per cent in sewage treatment plants in eastern European countries and by 40 per cent in western European countries. We extrapolated retention estimates from Sweden to other western European countries and from Polish estimates to other eastern European countries. We made these assumptions in order not to overestimate the nitrogen load from the human population.

3 Nitrogen retention potential of atmospheric deposition in a scenario where drained wetlands in the Baltic Sea drainage basin would be restored.

Previous studies of wetlands as nitrogen sinks

The GIS-database does not indicate what type of wetland ecosystems there are in the region. Since 95–96 per cent of the natural wetlands in the former USSR and Europe are bogs or fens (Aselmann and Crutzen, 1989), we assumed that all wetlands identified in the database are either bogs or fens.

The relationship between nitrogen load and retention by wetlands was obtained from a number of studies (Table 3.2). Many studies have found nutrient uptake rates in bogs to be lower than in fens (for example, Malmer, 1962; Small, 1972; Tilton, 1977). Retention data of bogs and fens compiled by Mander and Mauring (1995) show up to 75 per cent higher retention values for fens. However, retention estimates for fens are based on the TIN-method (the total inorganic nitrogen method measures only inorganic fractions) and

Table 3.2 *Literature Compilation of Nitrogen Load-retention Measurements*

Wetland type	Measuring method	Load ($kg\ km^{-2}\ y^{-1}$)	Retention ($kg\ km^{-2}\ y^{-1}$)	Time	Reference
Bog	TN	420	420	year	Rosswall and Granhall 1980[1]
	TN	1200	630	year	Verry and Timmons 1982[1]
	TN	1530	950	year	Horner 1986[2]
	TN	2000	1200–1400	year	Aerts et al 1992[1]
	TN	11,000	8800	year	Burke 1975[2,3]
	TN	13,000	5300	year	Slapokas 1991[1]
	TN	30,000	8100	year	Slapokas 1991[1]
	TN	32,000	31,000	year	Slapokas 1991[1]
Fen	TIN	9130	4880	Apr–Nov	Kadlec 1979[2,3]
	TIN	6570	3850	March–Nov	Kadlec and Tilton 1978[2,3]
	TIN	9130	3480	June–Nov	Kadlec 1981[2,3]
	TIN	6210	2380	June–Nov	Kadlec 1980[2,3]
	TIN	1900	765	May–Sept	Tilton and Kadlec 1979[2,3]
	TIN	1500	285	Aug–Oct	Kadlec and Tilton 1977[2,3]
	Interval:	**420–32,000**	**420–31,000**		

[1] After Leonardsson (1994)

[2] After Mander and Mauring (1992)

[3] Cited in Nichols (1983)

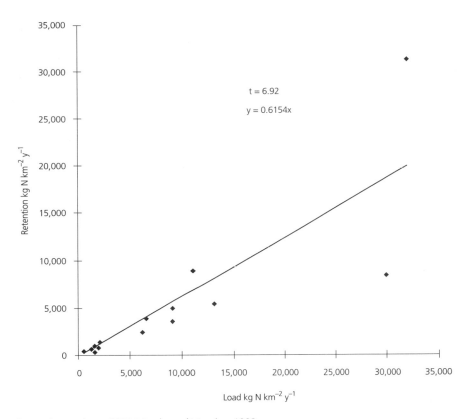

t = 6.92

y = 0.6154x

Source: Leonardsson, 1994; Mander and Mauring, 1992
Note: Y=0.6154x, n=14. The intercept has been set at the origin to avoid a leakage of nitrogen without adding load, as mass balances for both types of wetlands included in the study are positive or close to zero.

Figure 3.2 *Regression Line of Load-retention Relation*

for bogs on the TN-method (the total nitrogen method measures, all fractions of substances). Measurements based on a single fraction can be misleading due to reformulation of the fraction in question (Leonardsson, 1994). For example, the release of ammonia due to the decomposition of organic material or the formation of nitrate through the oxidation of ammonia. Also, the bog investigations are whole-year studies, while none of the fen investigations exceed nine months. Although there is some retention all year around (Mitsch, University of Ohio, personal communication) we assumed the retention for the time period not included in the fen studies to be zero.

The estimated load/retention regression line derived from these studies is presented in Figure 3.2.

The regression line helps us to obtain quantitative estimates of the effects of various nutrient loading. It is derived from empirical information from various sources, with consequent uncertainties associated with them. To capture some of the uncertainty in the regression we varied the slope by plus or minus 1 standard deviation (stdev) (1 stdev = 0.008891). A change by 1 stdev results in a 10–15 per cent change in the retention results.

Wetlands located high up in a drainage basin are often thought of as sources rather than sinks. In Sweden, for example, 9.5 per cent of the nitrogen load to the Baltic Sea originates from wetlands, and 88 per cent of that 9.5 per cent are from wetlands located north of the Baltic Proper (Löfgren and Olsson, 1990) in areas with low population and agricultural density. We have not been able to distinguish between sources and sinks as it would require much more specific and detailed information than is presently available for the region as a whole. Instead, we have estimated a range of retention capacities. The load in the scenario 1 is a conservative estimate including only deposition of atmospheric nitrogen to wetlands in the drainage basin. This is the minimum load that all wetlands actually receive although it varies between the grids. It includes nitrogen emissions to the air from human activities like industry, transportation, and livestock farming. In scenario 2 we add loads of nitrogen in processed sewage water from the human population. It is difficult to judge how realistic this scenario is. In one sense it is an overestimate as we assume that all processed sewage water with excretory is discharged into wetlands. On the other hand, it is an underestimate since we only account for the proportion of human excretory release that remains in processed sewage water, and we also do not account for leakage from forests and agricultural land.

We set the maximum load allowed to enter the wetlands to 32,000 kg N $km^{-2}y^{-1}$. The reason is that this was the highest load of retention in bogs and fens reported in the literature. We set this limit since we do not know the shape of the regression line above this load. Fleischer et al, (1991) have argued that the higher the load the higher the nitrogen retention in wetlands, on an absolute scale. Mander and Mauring (1995) found the regression between load and retention to be linear in wetland ecosystems. They included load levels as high as 1,000,000 kg km^{-2} y^{-1}, but their regression includes several other types of wetland. Even though there seems to be no absolute limit to retention (Fleischer et al, 1991; Mander and Mauring, 1995), it may not be very effective to use wetlands to retain or eliminate high concentrations of nutrients. Although wetlands can be and are used to process human wastewater and are (Etnier and Guterstam, 1991; Ewel, 1997), sewage treatment plants will always be needed in more densely populated areas. The comparative advantage of using wetlands as nutrient sinks lies in their capacity to remove diffuse nutrient runoff at low concentrations (Howard-Williams, 1985).

Calculating load and retention

Total load and retention were calculated using a grid-cell-based GIS. The study area was divided into 50x50 km grids with information on the distribution of wetland area and population in each specific grid. The Baltic Sea drainage basin (the Baltic Sea excluded) consists of about 700 grid cells. In scenario 1 we assumed that atmospheric nitrogen deposition per km^2 is the same within each grid cell. By placing total atmospheric deposition estimates per grid cell in the regression equation (Figure 3.2), we estimated the atmospheric nitrogen retention by wetlands for all grid cells. Total atmospheric deposition was estimated at 16,000–55,000 tons of nitrogen per year. The atmospheric nitrogen load entering wetlands per grid in the Baltic Sea drainage basin is shown in

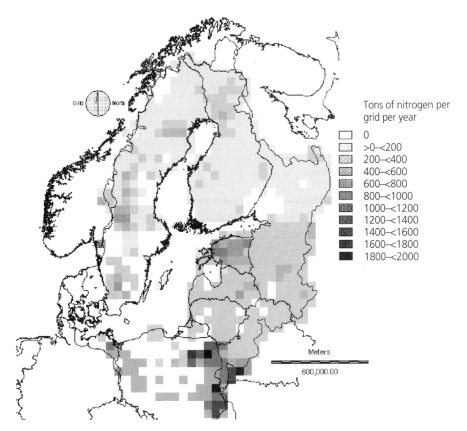

Tons of nitrogen per
grid per year

☐ 0
▨ >0–<200
▨ 200–<400
▨ 400–<600
▨ 600–<800
▨ 800–<1000
▨ 1000–<1200
■ 1200–<1400
■ 1400–<1600
■ 1600–<1800
■ 1800–<2000

Meters

600,000.00

Source: Map derived from the GIS database created by Sweitzer et al (1996)
Note: Each square represents an area of 50x50 km.

Figure 3.3 *High Range of Atmospheric Nitrogen Deposition Entering*
Wetlands

Figure 3.3. Nitrogen retention of atmospheric deposition in natural wetlands
was estimated to be 34,000–99,000 tons of nitrogen per year.

In scenario 2 we estimated total nitrogen load from the population and its
retention in each grid cell. By adding loads from all grids, we could estimate
total nitrogen load from the population entering the wetlands. We assumed
that all water containing nitrogen from human excretory waste discharged
from sewage treatment plants was processed in wetlands. Furthermore, we
assumed that nitrogen load from a population in a grid would be processed by
wetlands located in the same grid. In order to avoid situations where the
population in densely populated grids would discharge into very small
wetland areas, we set an upper load limit at 32,000 kg N km^{-2} based on the
maximum value included in our regression.

By using the regression equation we estimated the total nitrogen retention for the nitrogen in human excretory waste, after it has been processed

in sewage treatment plants. Finally we added the estimated total atmospheric nitrogen retention to the load from the population. Total load from the population in the basin was estimated at 255,000 tons of nitrogen per year. The geographical distribution of the excretory release of nitrogen in processed sewage water from the human population in the drainage basin is shown in Figure 3.4. Adding the nitrogen from the population in processed sewage water, the nitrogen retention was estimated at 57,000–145,000 tons of nitrogen per year.

In scenario 3 we estimated the retention potential of restorating previously drained wetland areas, and added that area to the present wetland areas. We only accounted for atmospheric deposition as a source of nitrogen, since the geographical location of potentially restored wetlands in relation to the human population, is unknown. Data on the percentage of drained wetlands could only be identified for Denmark, Estonia, Finland, Latvia, Poland and Sweden (Table 3.1). Statistics on drained wetland areas were available only at country level. As we do not know the geographical location of drained wetlands, we assumed that they would receive nitrogen loads per unit area corresponding to the average atmospheric load of the country in question. For Swedish wetlands, we assumed that they would receive an atmospheric deposition corresponding to the average deposition in the south of Sweden, since 90 per cent of all wetlands that have been drained were located in this area (Rydlöv et al, 1991). The nitrogen retention in this scenario was estimated at 196,000–261,000 tons of nitrogen per year.

The retention results in context

Total nitrogen input from the atmosphere and from land, entering the Baltic Sea throughout the drainage basin, has been estimated at approximately 1,170,000 tons per year, based on average emission data between 1970–1990 (A K Hallin, Department of Systems Ecology, Stockholm University, personal communication). Hence, existing natural wetlands in the drainage basin have the potential to retain an amount of nitrogen corresponding to 5–13 per cent of the total annual nitrogen emissions to the Baltic Sea. If drained wetland areas were to be restored, this capacity could correspond to as much as 18–24 per cent of the total annual emissions entering the Baltic Sea.

Other substantial nitrogen sources are leakages from forests and agricultural areas. We estimated that the nitrogen leakage from forests was 104,000 tons of nitrogen per year (approximately 100–235 kg km² per year based on Löfgren and Olsson (1990)) and that from arable land was 807,000 tons of nitrogen per year (estimated at 400–5000 kg km² per year (Elofsson (1997)). The amount of this leakage that reaches the Baltic Sea is unknown. We could not include these sources in our analysis due to lack of information on basic hydrology and the topographical location of both sources (forests and agricultural areas) and sinks (wetlands). If they were to be included, the amount of annually retained nitrogen by wetlands would presumably increase. The analysis highlights that we need to better understand the mosaic of ecosystems in drainage basins and how they affect and are affected by human activities.

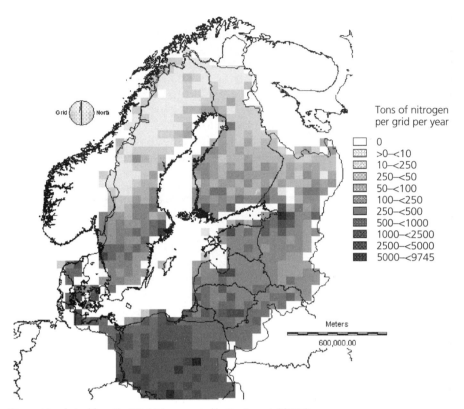

Tons of nitrogen
per grid per year

☐   0
▨   >0–<10
▨   10–<250
▨   250–<50
▨   50–<100
▨   100–<250
▨   250–<500
▨   500–<1000
▨   1000–<2500
▨   2500–<5000
■   5000–<9745

Meters

600,000.00

Source: Map derived from the GIS database created by Sweitzer et al (1996)
Note: Each square represents an area of 50x50 km.

Figure 3.4 *Distribution of Anthropogenic Direct Excretory Discharge*

THE WETLAND FOOTPRINT OF BALTIC CITIES

Several analyses of ecological footprints have been performed recently (for example, Folke et al, 1997; Larsson et al, 1994; Rees and Wackernagel, 1994). Here, we review our analysis of the wetland area appropriated for the retention of nitrogen from the 29 largest cities in the Baltic Sea drainage basin (Folke et al, 1997). These cities account for about 25 per cent of the human population within the Baltic Sea drainage basin. The ecological footprint is estimated by investigating the estimated areas of wetlands that would be needed to contain the nitrogen in the excretory waste released by humans. Consequently, it is an underestimate, as the emissions of nitrogen from food processing, household waste, car emissions, and other sources are not included in the analysis.

In the region, wetlands are increasingly used as a nitrogen-abatement technology, as there are indications that they serve as cost-effective filters for both point source and non-point source pollution (Gren 1995b). Again, we assume that all nitrogen releases from urban areas pass through sewage treatment plants, and that nitrogen remaining in the purified water is processed in wetlands. The data for our estimates is based on the analysis described in the previous section of this chapter, where the nitrogen-retention capacity of natural wetlands range from 0.4–1.1 tons of nitrogen per km². The results indicate that if nitrogen released from human waste from city inhabitants were to be processed in wetlands, 45–120 per cent of the presently available wetland area would be appropriated. This area is about 30–75 times larger than the area of the cities themselves. The analysis illuminates the often unperceived human requirements for ecosystem work, in this case the work of wetlands as waste processors. This work is 'hidden' because it has no value in the economy and people and policies seldom perceive it. Nevertheless, it is real.

SUMMARY OF MAJOR RESULTS

We have estimated the nitrogen-retention capacity of natural wetlands in the large scale Baltic Sea drainage basin, using and developing a wetland GIS database, statistics and other information from a variety of sources. The GIS database tells us that there are approximately 138,000 km² of wetlands (bogs and fens) in the Baltic Sea drainage basin, corresponding to 8 per cent of the whole area. The input of nitrogen into natural wetlands from atmospheric deposition (wet and dry) was estimated to be between 55,000–161,000 tons per year. Nitrogen in the excretory waste from the human population in processed sewage water was estimated to be 255,000 tons per year. There may also be leakage from forests and agricultural land into the wetlands, but lack of data on hydrology, topography and ecological information made it necessary to exclude this leakage in the analysis.

The capacity of the wetlands to retain the atmospheric deposition of nitrogen was estimated to be about 35,000–100,000 tons per year, and when the nitrogen input from the human population was added, the retention by

Table 3.3 *Wetland Ecosystem Area Appropriation by the 29 Largest Cities in the Baltic Sea Drainage Basin*

Country	No of cities	Appropriated ecosystem area (km² x 1000) Nitrogen retention
Czech Republic	2	1.8–4.9
Denmark	2	3.6–9.8
Estonia	1	1.4–4.0
Finland	1	1.1–3.0
Germany	2	1.3–3.5
Latvia	1	2.6–7.2
Lithuania	2	2.9–8.1
Poland	13	23.1–63.7
Russian Federation	1	14.7–40.4
Sweden	3	5.7–15.6
Ukraine	1	2.4–6.5
Total	**29**	**60.6–166.7**

wetlands was estimated to be between 55,000–145,000 tons per year. If drained wetlands were to be restored and their area added to the present wetland area, it was estimated that the nitrogen retention capacity would increase to between 195,000–260,000 tons per year. The results indicate that existing natural wetlands in the Baltic Sea drainage basin annually can retain an amount of nitrogen which corresponds to about 5–13 per cent of the total (natural and anthropogenic) nitrogen emissions entering the Baltic Sea every year.

Twenty-five per cent of the human population in the drainage basin lives in 29 major cities. To process nitrogen in human excretory waste after it has been treated in sewage treatment plants, the cities would require a wetland area – a wetland footprint – that is about 30–75 times larger than the area of the cities themselves (Table 3.3). It seems as if the retention service of wetlands accounts for a substantial nitrogen removal not only in specific local ecosystems, but also at the regional scale.

Part II
ECONOMIC AND ECOLOGICAL EVALUATION

4 COST-EFFECTIVE NUTRIENT REDUCTIONS TO THE BALTIC SEA

Ing-Marie Gren

Cost-effectiveness can be defined as obtaining one or several production targets at minimum cost. In our case, the policy objective corresponds to specified reductions in the nutrient load to the Baltic Sea. A condition for cost-effectiveness is that the marginal costs of all possible abatement measures are equal. Marginal cost is defined as the increase in costs when, in this context, a nutrient load to the Baltic Sea is decreased by 1 kg nitrogen or phosphorous. As long as the marginal costs are not equal it is always possible to obtain the same level of nutrient reductions at a lower cost by reducing the load via measures with relatively low costs. Thus, in order to calculate cost-effective nutrient reductions to the Baltic Sea we have to firstly identify all possible reduction options and their locations, secondly, quantify their reduction effect on nutrient loads to the Baltic Sea, and finally, calculate marginal costs for all options.

Since the 1960s, a large literature on cost-effective pollutant reductions to water bodies has accumulated. Generally, the focus has been on one polluting sector, and a small-scale watershed. (Kneese and Bower, 1968). The linkage of economic activities with pollutant discharge and deposition in water bodies has mainly been carried out for agricultural pollution (see Russell and Shogren, 1993, for an overview). The cost-effectiveness analysis carried out here thus differs from previous studies in terms of the scale of the study and also in that it includes the effluent outputs of several polluting sectors.

The identification of measures to reduce nutrient loads is based on Table 2.3 in Chapter 2, where the nutrient load data are presented. The measures will then involve the agricultural sector, sewage treatment plants, traffic and other nitrogen oxides emission sources. Table 2.3 further reveals that measures located in Poland will probably play an important role since the load of nitrogen and phosphorous in this country account for about 33 per cent and 50 per cent of the respective total loads. However, the choice of an efficient set of measures will be conditioned by their marginal costs.

As a first step in the cost-effective nutrient-reduction calculation, we will discuss the concept of marginal costs in the context of nutrient reductions to the Baltic Sea and present the calculated marginal cost of measures involving the sources listed in the nutrient load table. We have only included the countries within the drainage basin that have coastal zones coincident with the Baltic Sea. Wetland restoration is also considered as part of the options set. Only the coastal countries are involved in the Baltic Sea action programmes and negotiations over nutrients in the coastal waters. Then, we present the relationship between possible nutrient reduction targets and associated minimum costs for their achievement. It should be noted that the estimated marginal costs, and hence total minimum costs, are sensitive to the underlying assumptions. We will therefore present the results derived from a sensitivity analysis of the basic assumptions.

MARGINAL COSTS OF NUTRIENT REDUCTION MEASURES

The cost of obtaining a marginal reduction of the nutrient load to the coastal waters is determined by the cost of the measure in question and the amount of required load reduction. The prices and combinations of production factors such as labour and capital determines the cost of the measure. In general, for a given measure, the costs are increasing as levels of abatement rise. For example, it is less costly to increase the nitrogen abatement capacity at sewage treatment plants from a capacity of zero to, say, 10 per cent of incoming load, than to increase it from 50 per cent to 60 per cent. This implies that the marginal costs are increasing for higher reduction levels. The reduction at the coast achieved by the imposition of a given abatement method, say 1 kg nitrogen or phosphorous, is also affected by the location at which the measure is implemented. For example, a reduction by 1 kg at a source located by the coast implies 1 kg nitrogen or phosphorous reduction in the Baltic Sea. However, if the source is remotely located in the drainage basin some distance from the coastal waters of the Baltic Sea, only a fraction of the effect of the 1 kg reduction at the source is felt at the coast. The share of the reduction impact at the source that reaches the coast depends on how much retention of the nutrient occurs as the nutrient travels from the source to the coast. The higher the retention, the lower the impact is on the coast. This implies that, for a given marginal cost at the source, the marginal cost of reducing the load to the coast is higher for remote sources with consequent low impact on the coast.

The measures included in this study can be divided in to three different types:

1   reductions in the deposition in the Baltic Sea and on land within the drainage basin;
2   changing land uses to reduce the leaching of nutrients; and
3   the creation of nutrient sinks that reduce the transport (increased retention) of nutrients to the Baltic Sea.

The first class of measures include sewage treatment plants, reduction in air borne emissions from traffic and stationary combustion sources, and reductions in agriculturally-related deposition on land. Decreases in air emissions are obtained by the installation of catalysts in cars and ships, reductions in

Table 4.1 *Marginal Costs of Different Measures Reducing the Nitrogen to the Coast, SEK/kg Nitrogen Reduction*

Region	Agriculture	Sewage treatment plants	Atmospheric deposition	Wetlands
Sweden	20–242	24–72	135–9500	23
Finland	57–220	24–60	874–6187	66
Germany	20–122	24–60	210–3576	27
Denmark	23–200	24–60	544–3576	12
Poland	12–101	7–35	523–3412	10
Latvia	59–196	7–35	183–1195	20
Lithuania	72–208	7–35	254–1723	15
Estonia	55–192	7–35	153–1999	36
St Petersburg	43–236	7–35	353–1884	51
Kaliningrad	28–210	7–35	273–1593	43

the use of motor fuel and other petroleum products, and installation of cleaning technologies in stationary combustion sources. Reductions in the agricultural deposition of nutrients is achieved by the decreased use of fertilizers and reductions in livestock numbers, or livestock intensities. Another measure included is a change in the spreading time for manure from autumn to spring. Decreases in leaching from arable land are obtained by increasing the area covered by catch crops,[1] energy forests,[2] and ley grass. Nutrient sinks can also be created by constructing wetlands downstream in the drainage basin, close to the coastal water. In total, 13 different measures are included to facilitate reductions in nitrogen loads and 8 different measures can be aimed at phosphorous load reductions. The calculated marginal costs of nitrogen reductions for different types of measures and regions are presented in Table 4.1.

According to the results presented in Table 4.1, increased nitrogen-reduction capacity in sewage treatment plants is a low-cost measure in all countries. We also find low-cost measures in the agricultural sector – reductions in the use of nitrogen fertilizers and the cultivation of catch crops provide low-cost measures – and through the construction of wetlands. Measures reducing air emissions are relatively expensive in all countries. The reason for this is that only a small fraction of the decrease in nitrogen oxides from a given measure affects the Baltic Sea. The marginal cost of phosphorous reductions are much higher as measured in SEK/kg phosphorous reduction, see Table 4.2.

We note that measures involving sewage treatment plants are also a low-cost option for reductions in the phosphorous load, while the restoration of wetlands is relatively expensive. Note, however, that some of the measures involving land-use changes and reductions in the nutrient leaching from manure imply decreases in both nitrogen and phosphorous. The marginal

---

1 Crops grown after the principal crop is harvested in autumn.
2 Certain tree species which can be harvested for use as bioenergy.

Table 4.2 *Marginal Costs of Phosphorous Reductions, SEK/kg Phosphorous Reduction*

Region	Agriculture	Sewage treatment plants	Wetlands
Sweden	155–6604	41–52	18232
Finland	225–6080	41–52	1748
Denmark	144–2610	41–68	1202
Germany	188–2964	41–68	899
Poland	114–2033	20–100	611
Estonia	282–5622	20–100	6090
Latvia	234–5662	20–100	1234
Lithuania	186–6696	20–100	964
St Petersburg	230–4314	20–100	823
Kaliningrad	338–4290	20–100	545

costs presented in the tables do not account for this combined effect provided by these particular measures.

MINIMUM COSTS OF NUTRIENT REDUCTIONS

Currently, the impacts on the ratio of nitrogen and phosphorous in marine waters to changes in the nutrient loads are unclear (see Chapter 5 for a thorough discussion). Therefore, minimum costs are calculated for three different scenarious: reductions in nitrogen only, phosphorus only, or both nitrogen and phosphorus. The minimum costs for various reductions in either nitrogen or phosphorus are illustrated in Figure 4.1, where squares and triangles denote the costs of nitrogen and phosphorous reductions respectively.

We can see from Figure 4.1 that the costs for reducing the load of nitrogen are much higher than the costs for corresponding decreases in phosphorous loads. One important reason for this difference is that a relatively large share of the sources of phosphorous – sewage treatment plants – are located at the coast which implies relatively low marginal abatement costs. Common to reductions for both nutrients is the relatively rapid increase in costs at reduction levels above 40 per cent. At the 50 per cent reduction level in total loads of nitrogen and phosphorus, the minimum costs are approximately SEK 12 million and SEK 3 million respectively.

Several of the measures such as livestock reductions, changes in manure spreading time and wetlands restoration, imply reductions in both nitrogen and phosphorous. When one of these measures is implemented to reduce the load of one of the nutrients, reductions are obtained for the other nutrient 'free of charge'. This combined impact on several nutrients implies that these measures are relatively less costly for simultaneous reductions in nitrogen and phosphorus, than when these nutrients are considered separately. The differences in costs for these two strategies are shown in Figure 4.2.

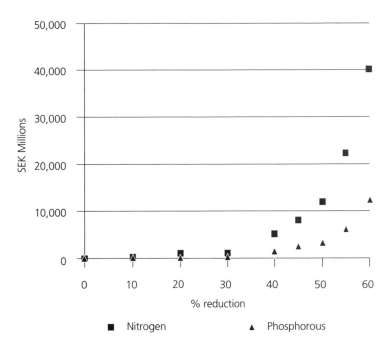

Figure 4.1 *Minimum Costs of Nitrogen and Phosphorus Reductions*

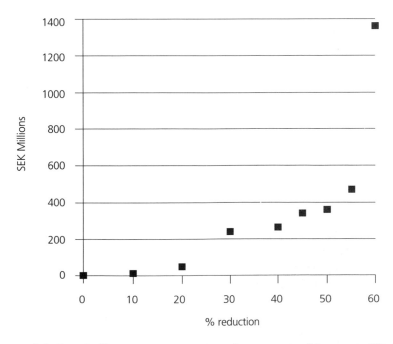

Figure 4.2 *Cost Difference Between Simultaneous and Separate Nitrogen and Phosphorous Reduction*

The difference in costs is relatively small for reduction levels below 40 per cent. This is because the total costs of separate reductions are also relatively small at these levels. Since the marginal costs for both nutrients increase rapidly at reduction levels above 40 per cent, the cost differences are much higher at these levels. At a 50 per cent reduction level the difference is approximately SEK 2.3 million per year.

Allocation of Cost-effective Measures

Given the information on nutrient loads and the costs of marginal nutrient reductions, we would expect that increased capacity at sewage treatment plants would play a significant role for both nitrogen and phosphorous reductions. The relative importance of this measure as compared to other measures is, however, dependent on its marginal cost in relation to other measures and the limits to its use. It may very well be the case that the capacity of sewage treatment plants is sufficient to achieve moderate nutrient-reduction targets. At higher reduction targets, such as the ministerial agreement of 50 per cent nutrient reduction, additional measures will also be required. In Figure 4.3, the cost-effective nitrogen and phosphorus reductions involving each type of measure for the 50 per cent overall reduction target are presented.

The cost-effective allocation of measures reveals that sewage treatment plants in the entire Baltic Sea drainage basin account for about 33 per cent of the total nitrogen reduction and 80 per cent of the phosphorus reduction. Wetland restoration is also part of possible options that contribute almost 33 per cent of the total nitrogen reduction and about 15 per cent of

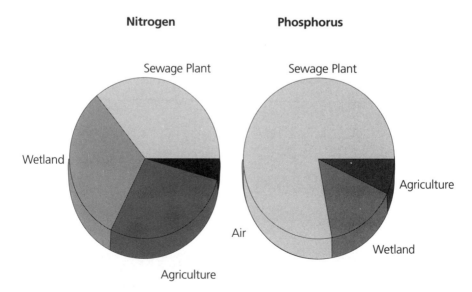

Figure 4.3 *Allocation of Measures at Cost-effective Nitrogen and Phosphorus Reductions by 50 Per Cent*

the phosphorus reduction. The agricultural sector contributes mainly by reductions in nitrogen fertilizers, the cultivation of cover crops, and changed practices for manure treatment. It can also be seen that measures directed towards reductions in air emissions are of less significance. This is due to their relatively low impact on the Baltic Sea, which increases the marginal costs of nitrogen load reductions. However, when we investigate instead the relative costs of the different types of measures at the 50 per cent target levels, we notice that the agricultural-change measures are relatively costly for nitrogen reductions. Construction of wetlands, on the other hand, will demand the largest share of abatement funding for a cost-effective phosphorus reduction of 50 per cent. This can be seen in Figure 4.4. Our expectations concerning the role of Poland in cost-effective nutrient reductions are also confirmed, see Figure 4.5.

From Figure 4.5 it is obvious that Poland will play an important role in both nitrogen and phosphorus reductions. In a cost-effective solution, the Polish reductions correspond to 50 per cent of the total nitrogen reduction and 60 per cent of the total phosphorus reduction. We can also see that Poland, together with the Russian Federation and the Baltic states, account for 75 per cent of the total reductions of both nitrogen and phosphorus. The results thus reveal that, in order to achieve cost-effective nutrient reductions, efforts should be concentrated mainly in Poland, but also in the Russian Federation, Estonia, Latvia and Lithuania. However, there might be 'local' reasons for the reduction of nutrient loads in specific coastal regions in

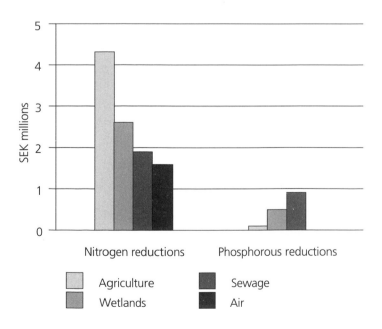

Figure 4.4 *Expenditure on Different Measures at Cost-effective Nitrogen and Phosphorus Reductions*

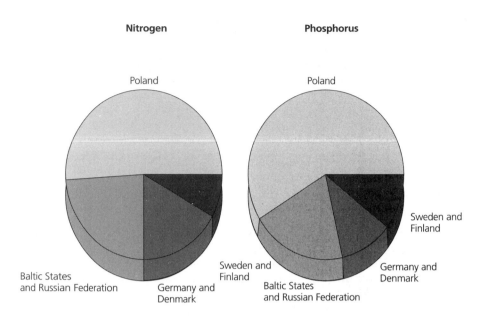

Figure 4.5 *Allocation of Cost-effective 50 Per Cent Nitrogen and Phosphorus Reductions in Different Regions*

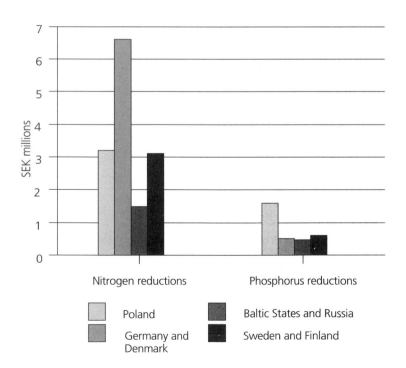

Figure 4.6 *Allocation of Costs for Different Countries at the Cost-effective 50 Per Cent Nitrogen and Phospohorus Reductions*

order to improve regional/local conditions. These aspects are not considered here.

The allocation of costs across the different countries reveals that the highest burdens are carried by Poland, Latvia, Lithuania, Estonia, and Russia (Figure 4.6).

The costs incurred by Poland, Estonia, the Russian Federation, Lithuania and Latvia account for about 80 per cent of the total costs of nitrogen reductions and approximately 65 per cent of the total costs of phosphorus reductions. An obvious question to ask at this stage is are these countries really willing to make such sacrifices without any type of compensation? This issue is discussed in detail in Chapters 7 and 8. However, adjustments within the countries concerned to the implementation of the measures can lower the direct costs (Figure 4.6). Analysis in Chapter 9 also indicates that the net impacts on the gross national productions of Poland, Lithuania, Latvia and Estonia, are in fact, relatively large.

SENSITIVITY ANALYSIS

In Chapter 2 it was pointed out that many assumptions underlying the estimates of nutrient loads in the Baltic Sea are somewhat simplistic and that some of the required data is either missing or of questionable reliability. These assumptions concern, in particular, the retention and leaching of nitrogen and phosphorus. These are of importance for the cost estimates, as they are used to predict the nutrient-reduction impact on the Baltic Sea from all the abatement measures used. For example, how would the total cost change if the leaching rate is increased or if retention is decreased, both of which imply that the impact on the Baltic Sea will be higher for all measures located upstream in the drainage basins. Likewise, there are assumptions concerning the internal capacity limit of individual measures. In the case of wetlands, for example, these limits are assumed to be the feasible area of restored wetlands. It is further assumed that sewage treatment plants cannot provide a reduction of more than the total nitrogen load from households and industry, corresponding to a 60 per cent total reduction.

In this chapter we calculate the cost impact of the 50 per cent nitrogen and phosphorus load reduction policy if changes in three different types of assumptions, biological, physical and economic, are made. The first type of change refers to alterations in assumptions about nutrient retention rates, wetland nutrient sink capacity and changes in the leaching from agricultural land use. The second type of change concerns the determination of the physical limits for the use of different measures, such as the maximum capacity of sewage treatment plants, the availability of land for wetland restoration and for changing agricultural practices. The third type of assumption concerns the costs to firms and households of reducing nutrient emissions.

Decreasing the rate of nutrient retention, increasing wetland nitrogen-purification capacity and the leaching rate due to agricultural land-use changes, all imply decreases in total minimum cost as compared to the base case, the results of which were presented earlier. The increased capacity of low-cost options has the same impact on total costs. If, for example,

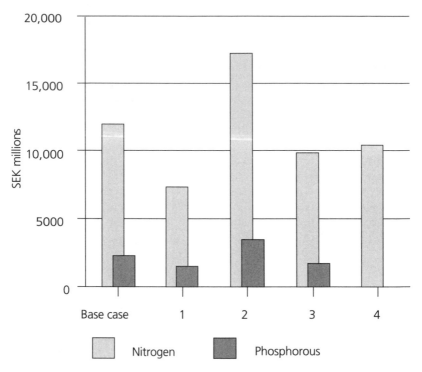

1 nutrient retention in Poland, the Russian Federation, and the Baltic States is reduced from 0.8 to 0.6;
2 the base case wetland retention capacity is reduced by 25 per cent;
3 the base case sewage treatment cleaning capacity is increased by 25 per cent; and
4 the costs of all air reduction measures are reduced by 50 per cent.

Figure 4.7 *Changes in Minimum Costs for 50 Per Cent Nitrogen and Phosphorus Reductions with Alternative Assumptions*

the area of land available for low-cost measures increases, the total cost of a 50 per cent reduction policy are reduced, since high marginal-cost measures are avoided. The impact of changing assumptions concerning cost estimates depends on the size of the sector with respect to nutrient loads and on the cost estimates in the base case. For example, if the sector is small and the costs are relatively high, we would not expect changing cost estimates to have a considerable impact on total costs. In Figure 4.7 we present the costs of a 50 per cent nitrogen reduction for changes to the assumptions applied in the base case.

According to the results, a decreased wetland nitrogen sink capacity of 25 per cent implies a doubling of the total costs for phosphorus reductions, as compared to the base case. A decrease in the rate of retention in the Baltic States, the Russian Federation and Poland from 0.8 to 0.6, reduces the total costs of both nutrient reductions by approximately one-half. We also see that the total cost of nitrogen reduction decreases by approximately 20 per cent if, in the base case, the cost of measures reducing air deposition have been overestimated by 100 per cent (scenario 4 in Figure 4.7). In general, changes to economic assumptions do not seem to have as great an impact on total costs as changes to biological and physical assumptions.

All of the sensitivity test results reported above assume, of course, the imposition of the overall 50 per cent nutrient reduction target. However, it may very well be the case that the costs are also sensitive to changes in the overall reduction target. Remember from Figure 4.1 the sharp increase in cost at levels exceeding 40–45 per cent reduction requirements, in particular, for reductions in the load of nitrogen. This increase is explained by the need to use relatively expensive measures once the 40 per cent overall re-

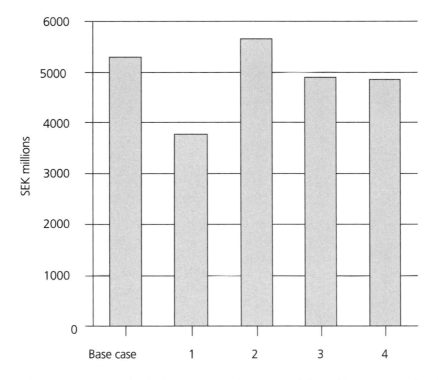

1 nutrient retention in Poland, the Russian Federation, and the Baltic states is reduced from 0.8 to 0.6;
2 the base case wetland retention capacity is reduced by 25 per cent;
3 the base case sewage treatment cleaning capacity is increased by 25 per cent; and
4 costs of all air reduction measures are reduced by 50 per cent.

Figure 4.8 *Costs of Alternative Assumptions for Reducing Nitrogen Loads by 40 Per Cent*

duction level is passed. Less nitrogen reductions achieved through low-cost measures due to lower effectiveness or tighter capacity limits, means adding higher cost measures in order to hit the 50 per cent reduction target. Costs are therefore sensitive to assumptions concerning the nitrogen reduction capabilities of low-cost options. For less ambitious nitrogen-reduction

Table 4.3 *Allocation of Nutrient Reductions and Costs of a Cost-effective and Proportional 50 Per Cent Reduction of Nitrogen and Phosphorous*

Region	Nitrogen			Phosphorus		
	Optimal		Propor-tional	Optimal		Proportional
	SEK m	% reduc-tion	SEK m	SEK m	% reduc-tion	SEK m
Denmark	1007	46	1859	276	60	162
Estonia	423	54	303	8	10	531
Finland	1542	41	1657	477	32	1402
Germany	526	15	43,340	258	55	150
Latvia	1321	66	261	237	55	193
Lithuania	742	58	241	109	52	92
Poland	3221	59	1119	1625	58	559
Russian Federation	1637	57	709	134	65	50
Sweden	1535	42	1913	81	19	10,438
Total	**11,956**	**50**	**51,402**	**3205**	**50**	**13,577**

targets, we would therefore expect costs to be less sensitive to changes in assumptions.

Figure 4.8 illustrates that the minimum costs for a 40 per cent nitrogen reduction are less sensitive than the costs for a reduction of 50 per cent. At most, costs are changed by about 33 per cent, when the assumptions concerning nitrogen retention rates in the Baltic States, the Russian Federation and Poland are altered. This is explained by their large shares of the total nitrogen load and also because they have relatively low-cost options still available for exploitation.

By definition, a divergence from the cost-minimizing principle implies higher costs. These cost increases could be justified on other criteria, such as equity. One equity-based principle, for example, would be the imposition of the same nutrient load reduction for all countries, as measured in a percentage decrease from the nutrient loads of a certain base year. Table 4.3 presents total cost estimates and those for each country, under the two possible options; minimizing costs for a 50 per cent reduction in the load from all the Baltic Sea countries, and each country reducing only its own loads by 50 per cent.

The total costs of both nutrients increase by about four times when the decision is changed from cost minimization to proportional nutrient reductions in each country. The increase in costs for a 50 per cent nitrogen reduction is underlined by the German measures directed towards expensive reductions in air emissions. The high cost of proportional phosphorus reductions is due to the high costs in Sweden, where considerable phosphorus reduction has already been undertaken.

CONCLUSIONS

The main purpose of this part of the book has been to calculate cost-effective reductions in the loads of nitrogen and phosphorous to the Baltic Sea. The results indicate a relatively high difference in the costs for similar nitrogen and phosphorus reductions. At the 50 per cent reduction level, annual minimum costs for nitrogen and phosphorus amount to approximately SEK 12,000 million and SEK 3000 million respectively. This difference is partly explained by differences in nutrient loads to the Baltic Sea; approximately 730,000 tons of nitrogen and 37,000 tons of phosphorus. Another factor is the relatively large number of low-cost measures for phosphorus reductions that have high impacts on the Baltic Sea due to their coastal locations. The restoration of wetlands and the expansion of sewage treatment plants provide low-cost options for reductions in both nutrients. For nitrogen reductions, certain low-cost measures are also found within the agricultural sector, such as reductions in the use of nitrogen fertilizers and the cultivation of catch crops.

The allocation of desired nutrient reductions and costs between countries is determined by their domestic total nutrients load, the availability of different measures and associated marginal costs. Countries with relatively low marginal costs and high shares of total nutrient loads, such as Poland, will then account for a relatively large share of the total reduction requirement. Under a cost-effective nitrogen reduction of 50 per cent, Poland accounts for roughly 40 per cent of the total required nitrogen reduction. The role of Poland is accentuated under a cost-effective phosphorous reduction policy. The Polish reduction then corresponds to almost 66 per cent of the total reduction requirement.

It is important to note that the cost estimates are sensitive to several underlying assumptions, in particular, to those concerning nutrient loads and transports, and the choice of total or proportional reduction options. However, although the numerical results must be interpreted with great caution, five results seem to be significant:

1   large differences in costs between nitrogen and phosphorus for the same percentage reductions;
2   the cost of simultaneous reductions in both nitrogen and phosphorus loads is less than the cost of corresponding separate reductions;
3   rapid increase in costs at reduction targets exceeding 40–45 per cent reductions;
4   the relatively high sensitivity of cost estimates to changes in the underlying assumptions at large reduction targets – an effect which seems to decrease for less ambitious reduction targets; and
5   abatement costs for both nutrients increase considerably when there is a switch from a cost-minimizing principle to one based on proportional reductions in nutrient loads.

Before concluding, important limitations of cost estimation and analysis need to addressed. Neither dispersion of costs across sectors other than those directly involved, nor transaction costs are included. As demonstrated in Chapter 9, the inclusion of dispersion effects for nitrogen reductions in Poland, the Baltic states, Sweden and Finland, does not, however, affect total costs very much, although the structural impacts are large. The transaction costs, defined as the costs of implementing and enforcing policies for achieving cost-effective allocations, are much more difficult to identify and quantify. As there are large differences in institutions which affect environmental policies between the countries, the transaction costs are likely to differ (Eckerberg et al, 1996). Further, it is quite likely that transaction costs are higher for measures directed towards non-point sources than towards point sources, which may also be owned by the public, such as municipal sewage treatment plants. The inclusion of transaction costs would thus affect, not only total costs, but also the cost-effective allocation of measures in different regions.

The exclusion of transaction costs implies an implicit cost-underestimation bias in this analysis. On the other hand, an overestimation effect may also be present due to the exclusion of other environmental net benefits associated with decreases in nutrient emissions. As mentioned in the introduction, such environmental benefits include improvements of the quality of inland waters and ground water, decreases in acidity in both soil and water, and reductions of the ground level ozone layer.

5 IMPACTS OF CHANGED NUTRIENT LOADS ON THE BALTIC SEA

Fredrik Wulff

In the introduction of this book it was illustrated that one important aim of this study was to compare the costs of nutrient reduction with associated benefits. In Chapters 2, 3 and 4 we have assembled and analysed sufficient data to arrive at the cost estimates. These measure the cost of reducing the nutrient load to coastal waters. But what happens then? How are the nutrients transported in the Baltic Sea and how do changes during transportation affect the biological/ecological conditions of the Sea? These are necessary questions to be answered in order to compare nutrient-reduction costs with benefits. However, the benefits measured in monetary terms are, in practice, as difficult to obtain as the cost estimates of nutrient reductions.

Before any benefit measurements can be accomplished at all we need to understand the biological/ecological impacts of changes in the nutrient loads, for example, which species are affected and when the impacts occur. As will be evident from this chapter, we are currently very far from a complete empirical understanding of the relationship between nutrient loads and ecological functioning in the different basins of the Baltic Sea. We have, however, some qualitative understanding, as illustrated in Figure 5.1. All of these effects (and more) of eutrophication are very well documented from monitoring programmes and scientific studies.

Almost 25 per cent of the bottom of the Baltic Proper has permanent oxygen deficiency and no macroscopic life, due to the increased production of organic matter. In more shallow coastal regions, there is, on the other hand, more life than ever due to eutrophication. However, valuable fish species have been replaced by less valuable species, and massive algal blooms threaten the recreational value of beaches and archipelagos, etc. It is known that excess load of a limiting nutrient to a water system implies a change in the concentration ratio of the nutrient in question, which, in turn, leads to increased algae and phytoplankton production. As a result of this, oxygen deficits may occur that reduce the areas available for successful cod reproduction. On the other hand, other types of fish species may increase.

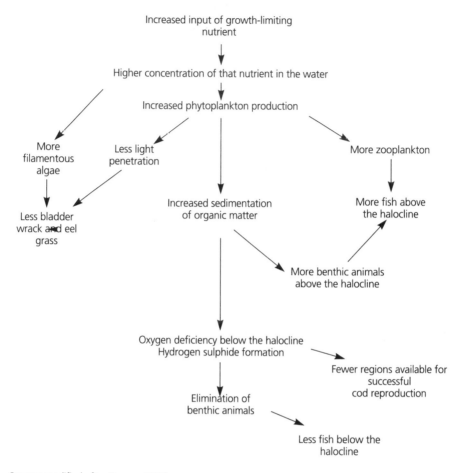

Source: modified after Bernes, 1980

Figure 5.1 *The Effects of Eutrophication on the Baltic Sea Ecosystem*

Obviously, the impact of changes in the supply of nutrients and algal growth is determined according to which nutrient, nitrogen or phosphorus, is limiting. This chapter therefore starts with a brief description of the role of the limiting nutrient. We then go on to examine the way in which the limiting nutrient differs between water basins, which has implications for nutrient reduction requirements as well. However, nutrient reductions to one sub-basin also have implications for other basins due to marine water streams. This is demonstrated in this chapter on both basin-wide and regional scales.

LIMITING NUTRIENT

For algae and other plants to grow and build up organic matter, sunlight, carbon dioxide and various nutrients are needed, primarily, nitrogen and phosphorous. These nutrients are required in certain proportions – for most aquatic organisms 16 nitrogen atoms are used for each atom of phosphorus. This ratio, 16:1, is typically found in both organisms and in concentrations of inorganic nutrients in most of the world's oceans. The adjustment to this so called 'Redfield ratio' in the sea occurs through two biological processes – nitrogen fixation and denitrification – that affect the inorganic nitrogen concentrations in opposite directions. However, this ratio is often far from 16:1 in many coastal marine regions.

In the sea and in lakes, cyanobacteria have the ability to convert inert nitrogen gas (N_2) into useful inorganic nutrient nutrients (ammonia (NH_3) or ammonium (NH_4)). Nitrogen fixation is a highly energy-demanding process that is only advantageous for bacteria to deploy in environments with excess phosphorus and little inorganic nitrogen available. For the Baltic Sea, nitrogen fixation by cyanobacteria (sometimes also called blue green algae) is an important process and contributes to about 15–20 per cent of the nitrogen inputs. However, it is only important in the Baltic Proper, south of the Åland Sea, where there is a large excess of phosphorus (a nitrogen:phosphorous ratio of about 4). In the northern basins the ratio is balanced (16 in the Bothnian Sea) or there is even an excess of nitrogen further north (40 or more in the Bothnian Bay). Thus, the primary production of organic matter is nitrogen limited in the Baltic Proper, but phosphorus limited in the Bothnian Bay. However, nitrogen fixation is apparently not sufficient to bring the nitrogen:phosphorous ratio to a balanced 16:1 in the Baltic Proper. Furthermore, the inputs to the Baltic have a very high ratio, about 40:1, thus bringing an excess of nitrogen to the Sea. Why is so much nitrogen lost, reducing the ratio to 4:1 or even less, and creating a nitrogen limited system with a large excess of phosphorus? This is due to another important microbial process, called denitrification: When organic matter is decomposed, oxygen is utilized by the bacteria and inorganic nutrients, ie, phosphorus and ammonia are released. Ammonia is further converted (oxidized) to nitrate (NO_3) if oxygen is available. If all the oxygen is consumed, the decomposition of organic matter continues, but now nitrate is used instead of oxygen by the bacteria. This denitrification process converts inorganic nitrogen into atmospheric gas. More than 80 per cent of the annual nitrogen inputs are lost via denitrification. Denitrification occurs primarily in the sediments and the shallow Baltic offers ideal conditions for this process.

Denitrification can thus be looked upon as a useful process, an 'ecological service' for reducing nitrogen loads and counteracting eutrophication. However, there is a downside. Although some of the phosphorus inputs are withdrawn from the biogeochemical cycles by adsorption to sediments, a larger fraction (than for nitrogen) remains in the water column and an excess of phosphorus is created. For various geochemical reasons the 'phosphorus trap' is more efficient in the Bothnian Bay. Thus, conditions are favourable for nitrogen-fixing cyanobacteria in the Baltic Proper. Blooms of cyanobacteria occur regularly there in late summer, floating to the surface and accumulating in foul-smelling beds along the shores. They are sometimes toxic and cause the death of cattle and dogs if they drink the water.

BASIN-WIDE IMPACTS OF CHANGES IN NUTRIENT LOADS

Although the impacts of greater nutrient inputs are well documented, the quantitative relationships between variations in the loads of nutrient and their concentrations are poorly understood. For instance, the eutrophying processes will alter the oxidation and reduction (redox) conditions. Thus, the biogeochemical pathways and efficiency of internal sinks of nitrogen and phosphorous, through denitrification and phosphorous adsorption, will be altered. The various sub-basins of the Baltic Sea have differences in water and nutrient residence times, load and internal biogeochemical processes that will alter the nitrogen:phosphorous ratios to a very high degree.

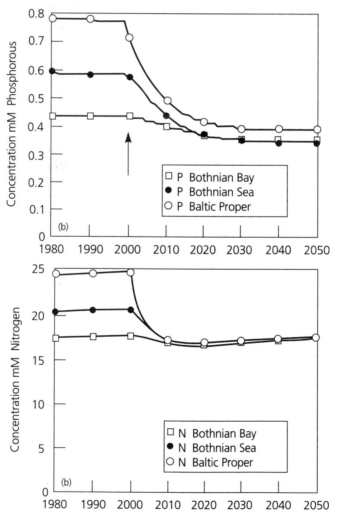

Figure 5.2 *Time Path of Concentration Ratios in Three Sub-basins from 50 Per Cent Reductions in Nitrogen and Phosphorus Load to the Baltic Proper*

In order to predict the impact of nutrient reductions in one sub-basin on another, the empirical budget model of the Baltic Proper has been expanded to cover all three sub-basins. The model consists of the three coupled basins with advective water and nutrient transports between these and with Kattegat. Empirical relationships between load, nutrient concentrations and advective transports from extensive data sets of the regions collected over a number of decades have been used to empirically calculate relationships between concentration and internal nutrient sink terms. These models rely on past relationships between inputs of water, nutrients and the observed trends in concentrations in the different basins of the Baltic. By means of these models different scenarios of nutrient reduction have been simulated.

Dynamics of nutrient-load reductions by 50 per cent

In this chapter we introduce the time perspective, disregarded in previous chapters. It turns out that a considerable time period is required to reach the new steady state concentration ratios from a given reduction in nitrogen and phosphorus loads. In Figure 5.2, the time path of concentration ratios from a 50 per cent reduction in both nitrogen and phosphorus loads to the Baltic Proper in the year 2000, is illustrated.

As can be seen from Figure 5.2, the nitrogen concentrations reach this new steady state within 10 years while it takes about 25 years for phosphorous. This is due to the inherently different behaviour of these nutrients in the Baltic as in most marine systems. Denitrification represents an efficient internal nutrient sink for nitrogen while phosphorus reduction is less efficient in this brackish system (Wulff et al, 1990). If nitrogen is the limiting nutrient of the Baltic Proper, we would thus expect a quicker response by the Baltic Sea to reductions in nitrogen than to phosphorus.

Nutrient concentration ratios and reductions in the load of nitrogen

If only the nitrogen load to the Baltic Proper is reduced, there are no effects on future phosphorous concentrations. Nitrogen will then decrease to lower levels than if both nitrogen and phosphorus are reduced (Figure 5.3).

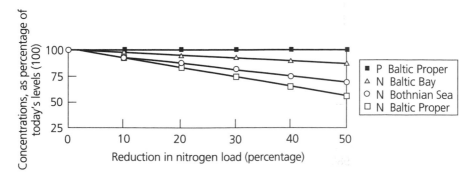

Figure 5.3 *Nutrient Concentration Ratios and Reductions in Nitrogen Load to the Baltic Proper*

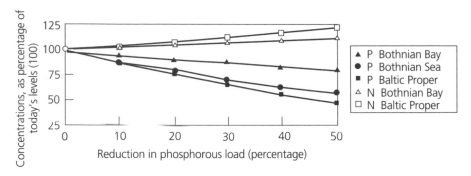

Figure 5.4 *Nutrient Concentration Ratios and Reductions in Phosphorus
to the Baltic Proper*

Within the modelling framework, more nitrogen is incorporated into organic matter and subsequently lost via denitrification if the phosphorous
levels are kept high. The model does not, however, account for the possible
increase in atmospheric nitrogen fixation by cyanobacteria that may occur at
reduced nitrogen:phosphorous ratios and high phosphorus concentrations.

Nutrient concentration ratios and reductions in the load of phosphorus

This scenario was included since, as was shown in Chapter 4, the phosphorus abatement costs are much less than for corresponding percentage
reductions in nitrogen. The simulation results are presented in Figure 5.4.
The model simulations show corresponding reductions of phosphorus concentrations to nitrogen (18 per cent for the Baltic Proper and 11 per cent for
the Bothnian Bay). A decrease of input and concentration of phosphorus
means that less nitrogen will be utilized in the biogeochemial cycles, since
these nutrients are utilized in fixed stochiometric Redfield ratios (16 moles
of nitrogen for each mole phosphorus). Thus, less nitrogen will be incorporated into organic matter and subsequently mineralized and denitrified
(lost) in this scenario. The Baltic Proper and Bothnian Sea will change
from nitrogen-limited to phosphorus-limited systems in this scenario.

Concentration ratios and reductions in both nitrogen and phosphorus

The impacts of reductions in both nutrients on the concentration ratios are
very similar to those of separate reductions (Figure 5.5).
    According to the results presented in Figure 5.4, the final concentration
of phosphorus and nitrogen are about 50 and 70 per cent of the levels of
today for the Baltic Proper. The coupling of the basins and the associated
transport of nutrients between them is shown by the percentage reduction
of nutrient concentrations in both the Bothnian Sea (59 per cent for phosphorus and 84 per cent for nitrogen); and the Bothnian Bay (81 per cent for
phosphorus and 98 per cent for nitrogen), despite the fact that there is no

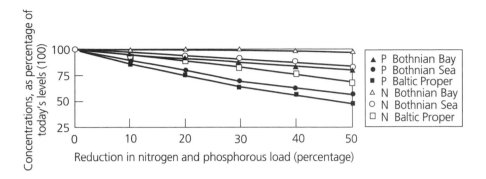

Figure 5.5 *Nutrient Concentration Ratios and Reductions in Both Nitrogen and Phosphorus to the Baltic Proper*

change in the inputs from the corresponding drainage basins. The relatively small effect on nitrogen levels in the Bothnian Bay is because primary production in this basin is phosphorus limited. Thus, there is an excess of nitrogen that is not incorporated into the biogeochemical cycles in this system. A phosphorus reduction in the Bothnian Bay will also further reduce primary production in this oligotrophic system and then further reduce microbial degradation and denitrification.

REGIONAL SCALE – GULF OF RIGA

The model simulations presented above illustrate the consequences of nutrient reductions on a basin-wide scale. However, decisions about abatement policies are often made on the basis of concerns over the local or regional environment and may produce outcomes that conflict with the protection of the whole basin. These conflicts, in a Baltic-wide perspective, are discussed by Wulff and Niemi (1992) and are further explored in a regional study on the Gulf of Riga (Yurkovskis et al, 1993). A nutrient budget was made that describes nutrient inputs, retention and exports of nitrogen and phosphorous from this highly eutrophicated bay adjacent to the Baltic Proper. This is a region of the Baltic where lack of sewage treatment contributes to a very large input of phosphorus to the sea. A reduction of phosphorus inputs of more than 30 per cent would occur if a modern sewage treatment plant were built for the city of Riga. A model was therefore built to explore the consequences of different nitrogen and phosphorus reduction schemes on the Gulf and the Baltic Sea. The results are presented in detail in Dahlberg et al (1996). One example of this exercise is illustrated in Figure 5.6.

Figure 5.6 illustrates the effect on the net exports of nitrogen and phosphorous to the Baltic in relation to different levels of phosphorus reduction in the inputs to the Gulf. Naturally, a phosphorus reduction in inputs will result in reduced exports of phosphorus. However, more nitrogen will then be exported to the (nitrogen-limited) Baltic Proper. The reason is that a decrease in the load of the limiting nutrient phosphorus decreases biological

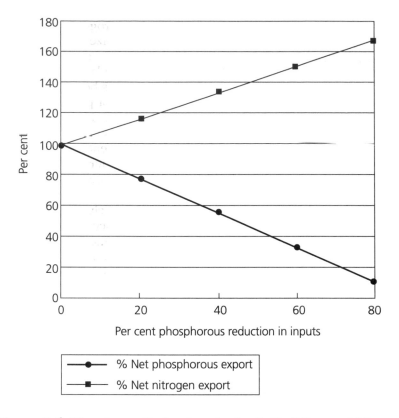

Figure 5.6 *Phosphorus Reductions in the Gulf of Riga and Nitrogen Exports to the Baltic Sea*

production, which, in turn, implies less use of nitrogen which is then exported to the nitrogen limited Baltic Proper. These nitrogen exports may then generate increased eutrophication damage in the Baltic Proper. It seems to be the case that by reducing phosphorus loads to the Gulf of Riga the eutrophication damages are exported to the Baltic Proper, while the condition of the Gulf will improve.

CONCLUSIONS

The simulations presented in this chapter, although based on very simplistic models, are built on empirical relationships and basic physical and biogeochemical properties. They show very clearly the basic features and interactions of the hydrodynamics and the biogeochemistry of nitrogen and phosphorous in the Baltic Sea region. The overall model clearly demonstrates that it is reduction of inputs to the Baltic Proper that is most efficient in reducing concentrations in this basin. A strategy where all inputs are reduced to the same degree is not optimal, since the situation in the two northern basins is not critical in terms of eutrophication (and only small amounts are exported southwards).

These simulations also demonstrate that both nitrogen and phosphorous inputs have to be reduced. This is also emphasized in the regional Gulf of Riga study where a phosphorus removal alone might actually increase the net export to the off-shore Baltic Proper. On a large-scale perspective, it is demonstrated that it will take several decades before the nutrient levels are returned to an acceptable level, particularly for phosphorous.

It is difficult to estimate the ecological consequences from the output of this model alone. The results have to be related to the empirical knowledge of the state of the Baltic at different concentrations of nutrients. The 'new' nutrient concentration corresponds roughly to levels found during the 1960s, before the drastic deterioration of the Baltic environment occurred. We would expect less primary production of organic matter and thus less frequent periods of oxygen deficiency in the deep basins. It is also likely that the decrease in phosphorus concentrations will reduce the frequency of cyanobacterial bloom during late summer. These are now favoured by the high phosphorus concentrations (and low nitrogen:phosphorous ratio) found during summer and cause accumulations of sometimes toxic algal mats on the surface of the Baltic Proper.

6 THE BENEFITS OF A LESS EUTROPHICATED BALTIC SEA

Tore Söderqvist

Does it really pay to reduce the load of nutrients to the Baltic Sea so that the conditions prevailing prior to the 1960s can be restored? Or, to put it differently, are the costs of a 50 per cent nutrient reduction – which should be enough to reach the biological conditions of before the 1960s – warranted by the benefits? From Chapter 5 we learnt that such reductions will change the concentration ratios of nitrogen and phosphorus. However, in order to find corresponding monetary measurements, this information is not enough. The reason is that people who make judgements over values cannot properly comprehend the role of concentration ratios without further information on impacts such as frequency of toxic algae blooms, changes in the composition of fish species, etc. An important challenge for the benefit-estimation exercise – human welfare gains or losses – is thus to 'translate' concentration ratios into biological functioning in a way that people can relate to and understand. In this chapter an attempt to approach this problem is described. Prior to the presentation of the specific benefits to the Baltic Sea, the general background to benefit estimation is reviewed.

JUDGEMENTS AND THE QUANTIFICATION OF ENVIRONMENTAL BENEFITS

Every human activity affects the environment in some way or another. The consequences may sometimes be negligible, but many activities result in significant environmental effects. For example, consider the construction of a new motorway through a forest. A list of environmental impacts would surely include the destruction of organisms such as trees and other plants. In addition, the motorway would probably, for some people, increase the attraction of going by car, meaning that traffic emissions would increase, other things being equal. On the other hand, transport would be faster and safer – traffic jams and accidents would be reduced. Other examples of projects that would probably affect the environment significantly are the construction of wastewater treatment plants, the restoration of wetlands and the introduction of incentives for farmers to reduce their use of fertilizers. Such actions would reduce the inflow of nutrients to surface waters and groundwater and may thus serve to reduce the effects of eutrophication.

A list of all the consequences – welfare gains and losses – of a given project can be used as a basis for a judgement about whether a project *should* be realized or not, given that one accepts a teleological perspective. Such a perspective implies that ethical judgements concerning an action are based on the *consequences* of an action, not on the action itself (Hanley and Milne, 1996). To go from an identification of consequences to an evaluation of them is, however, far from a trivial step, especially if some consequences are judged to be beneficial and others to be detrimental. Is the net effect positive or negative?

Economists typically approach this issue by characterizing the consequences as either benefits or costs. Since both benefits and costs are usually measured in monetary terms, it may be possible to compare them and to see whether the benefits of a project outweigh the costs or not. Needless to say, this is usually not easily done. Besides the fact that it may not be possible to identify all the consequences, some benefits and costs are difficult to measure. The economic approach also has an anthropocentric ethical basis. This means that values that possibly exist independent of humans are put aside and not directly considered (Turner, 1999). In some cases, the benefits and costs concern goods and services that are traded on well-functioning markets. Economists usually regard such benefits and costs as quite straightforward to quantify, as economic theory suggests that there are linkages between people's well-being and their market behaviour. When people are given a free choice between different commodities, they are supposed to make the choice about which is the best for their well-being for a given income. Economists' special liking for market data thus have grounds that are more fundamental than may be perceived at the first glance.

In the case of the project examples mentioned above, it is likely that market data can be used for calculating, for example, the construction costs for the motorway. Likewise, if agricultural yields reduced due to restrictions on fertilizer use, extra costs incurred by farmers may be estimated from market data on agricultural products. On the gains side, a reduced eutrophication of the Baltic Sea may result in a larger cod population and thus larger cod catches, for example. Data about the size of this particular gross benefit from reduced eutrophication may be obtained from the fish market.

Market data may thus help in assigning monetary values on to some benefits and costs. However, markets are often not well-functioning due to, for example, regulations and lack of information. Moreover, some people's free choices may be short-sighted so that the well-being of future generations are not taken into account. Another problem is that some commodities are nonmarket goods, ie, they are not traded in any market. For example, 'ecosystem services', ie, 'commodities' supplied to humans by the ecological systems, are typically untraded. Ecosystem services include, *inter alia*, life-support to humans by ecosystems' maintenance of the composition of the atmosphere, the climate, the soil and the hydrological cycle, as well as food, recreational opportunities, beauty (Daily, 1997).

The importance of ecosystem services for people's well-being implies that a change in their supply due to human activity somehow has to be included in the list of benefits and costs. To some extent, such an inclusion will probably always be qualitative. However, economists have developed methods that try to estimate at least some of the benefits of an improved environmental quality (or, analogously, some costs of environmental damage) in quantitative terms. See, for example, Freeman (1993) for a detailed presentation of these valuation methods. These 'valuation methods' suggest that people's willingness to make sacrifices (in dollars, grammes of gold, kilos of apples, or whatever) to obtain something reveals information on how they value it. The use of market data (when available) for measuring costs and benefits is based on precisely the same approach. If people refuse to consider trade-offs – though most of us make them, at least implicitly – this has to be taken into account outside a conventional economic analysis.

The valuation methods employ data on either people's observed market behaviour or their hypothetical market behaviour. Despite the fact that ecosystem services are typically nonmarket goods, data about *observed* market behaviour can be useful when there is some linkage between the ecological service and one or more market goods. For example, one usually has to pay something for getting to a beach, if one does not happen to live next door to it. Broadly speaking, such expenses can be regarded as the lowest value of the beach in question. If a visitor did not assign the beach a value that covers the transport costs, the visit would not have occurred. This simple observation is the point of departure for the travel cost method (TCM), one of the valuation methods that employ data about observed market behaviour.

Linkages between ecosystem services and market goods, if any, are sometimes weak or difficult to establish. Valuation methods based on *hypothetical* market behaviour try to overcome this problem by designing a market for the ecosystem service itself. The contingent valuation method (CVM) is the most popular method of this kind. A CVM application requires a description of a change in the supply of one or more ecosystem services. This description is communicated to a sample of individuals by the use of survey instruments such as mail questionnaires or interviews. Questions are then usually posed about the person's willingness to pay (WTP) for a realization of the change, but actual market transactions including payments do not take place (Bateman and Willis, 1999).

The use of these valuation methods allows the comparison of benefits and costs on a common monetary basis when changes in ecosystem services are involved. It may not be possible to quantify all benefits and costs, however. Incomplete information may, however, in some instances be sufficient for policy making. For example, this could be the case if it has been possible to estimate only a fraction of the total benefits provided by an environmental asset, but this fraction happens to be greater than the estimate of total costs.

CONTINGENT VALUATION OF THE BALTIC SEA

A number of empirical studies have been carried out in Lithuania, Poland and Sweden in order to obtain some information on the size of the benefits of a reduced eutrophication. Results from CVM applications in Lithuania, Poland and Sweden are available (see Chapter 7 in this volume; Söderqvist, 1996a; Turner et al, 1999). Sandström (1996) has employed the TCM for estimating recreational benefits from an improved water quality along the Swedish Baltic Sea coast.

The cost and benefit studies were coordinated by studying benefits in at least one transition economy (Poland) and at least one established market economy (Sweden). The data obtained allowed aggregate estimates for these two countries, but also more controversially for the two groups of economies around the Baltic Sea. It was also seen as desirable not to rely solely on methods that use either observed or hypothetical data. All the available methods have weaknesses, but in particular, the pros and cons of the CVM are subject to an intense discussion (Diamond and Hausman, 1994; Hanemann, 1994; Portney, 1994; Bateman and Willis, 1999). Its use of data on hypothetical market behaviour constitutes the core of this discussion. On the other hand, it is precisely this feature that allows the CVM to cover benefits that are difficult or even impossible to infer from observed market behaviour – the relationships between the ecosystem service in question and market goods may be weak or non-existent.

Consequently, both CVM and travel-cost studies were launched in Poland and Sweden. Recreational benefits are estimated by the latter studies, whereas the CVM studies have the potential to estimate total benefits from a reduced eutrophication of the Baltic Sea. That is, benefits that may be compared with the costs of the large-scale action plan analysed in Chapter 4. However, since data on benefits perceived by people are obtained by a CVM study, these benefits are 'total' only in a special sense. It is virtually impossible to communicate a detailed description of all the effects of a reduced eutrophication, which means that the perceived benefits from a reduced eutrophication may differ from the total ones. Another problem is the incomplete scientific knowledge of eutrophication and its effects. Moreover, to take a stand on the benefits of reduced eutrophication is far from an everyday situation for people, and a lack of familiarity may impair the working of the CVM approach.

Sweden

The main Swedish CVM survey was carried out in January–March 1995. A mail questionnaire was developed for the survey, see Söderqvist (1996a) for details. It included, *inter alia*, questions on how people use the Baltic Sea, information on eutrophication and its effects, questions on the knowledge and experience of the eutrophication effects, a description of the valuation scenario, and WTP questions. In the valuation scenario, the respondents to the questionnaire were asked to imagine the following:

1   A large-scale international action plan to combat the eutrophication has been proposed. If no actions are taken, there is an increased risk of more widespread and common eutrophication effects.
2   Cost-effective actions would be taken.

3   The plan would be financed by an extra environmental tax on households, farmers, firms, etc, introduced in all Baltic Sea countries and ear-
marked for the specific purpose.
4   The plan would imply that the eutrophication in 20 years has decreased
to a level that the Baltic Sea systems could sustain.

A sustainable level was interpreted to be a situation where the present ecological imbalances (carefully described in the questionnaire) had been rem-
edied. The 20-year perspective selected was consistent with scientists' predictions that a 50 per cent reduction of nutrient loads today would result in
10–25 years time in nutrient concentrations found in the Baltic Sea during the
1960s, before the rapid deterioration of the Baltic ecosystems (see Chapter 5
by Wulff). The selected payment vehicle (an extra environmental tax) was
found in the pilot work to be the one preferred by most people. A few examples of measures that the action plan would include were mentioned, but only
in very general terms. The objective here was to provide examples detailed
enough to make the action plan realistic, but also vague enough to make subjects think about benefits (B) from a reduced eutrophication rather than benefits minus action costs (B-C). If this objective was met, C as estimated by Gren
in Chapter 4, could be deducted from B in order to compute net benefits.
   The valuation scenario was followed by the following discrete choice
(DC) question:

> *'If there were a referendum in Sweden about whether to launch
> the action plan or not, would you vote FOR or AGAINST the
> action plan if your environmental tax would amount to SEK
> X per year during 20 years?'*

Four response choices were given in the question:

> *(1) I would vote for the action plan;*
> *(2) I would vote against the action plan;*
> *(3) I would return a blank ballot-paper; and*
> *(4) don't know.*

The survey population was defined as the 6,631,100 individuals who were
registered in Sweden and who were between 18 and 85 years of age on 16
January 1995. This questionnaire was distributed to 700 randomly sampled
Swedes between 18 and 85 years old. Seven different amounts X ('bids')
ranging from SEK 1000 to 25,000 were randomly distributed among the
sampled persons (only one amount per subject). In addition, subjects not
voting for the action plan were asked to also consider a zero tax.
   Some respondents' answers to the WTP question may reflect some protest against the valuation scenario. For example, if a respondent thinks the
payment vehicle is unacceptable, he may vote against the action plan even
if his WTP for the environmental improvement *per se* is considerable. Benefit estimates from a CVM survey are valid only for the policy prescribed in
the valuation scenario. This means that responses that reflect a protest
against the scenario should be identified and considered separately. This

was accomplished in this survey by a question about subjects' motives for
their answer to the WTP questions. An open-ended question format was chosen in order to avoid cues that would make subjects' opinions endogeneous.
The results of a categorization of these answers are reported in Söderqvist
(1998) and are used in the analysis below.

The response rate in the main survey was slightly more than 60 per cent,
which is a normal figure for Swedish CVM mail questionnaire surveys (see
Söderqvist, 1996b). As a brief indication of the kind of information that can be
obtained from the survey, it can be said that Swedes have, on average, a close
contact with the Baltic Sea. At least one of the following categories turned out
to be applicable to almost 70 per cent of the respondents: work related to the
Sea; main residence near the Sea; cottage near the Sea; holidays at the Sea in
1993 or 1994; leisure use of the Sea in 1993 or 1994. More than 90 per cent of
the respondents had heard about the eutrophication, and almost 85 per cent
about algal blooms, which was the most familiar eutrophication effect. Thirty
per cent said that they had some personal experience of eutrophication effects, and algal blooms were the most experienced effect. The results also
indicate that on average, respondents regarded the eutrophication of the Baltic Sea as an environmental problem of the same significance as, for example,
the acidification of land and lakes in Sweden. See Söderqvist (1996a) for an
analysis of how socio-economic and other background variables are related to
respondents' answers to the voting question.

The close contact with the Baltic Sea, the high awareness of the eutrophication and the recognition of eutrophication as a serious environmental
problem suggest that a relatively high willingness to vote 'For' the action
plan was to be expected. A majority of the respondents who were con-
fronted with bids ≤ SEK 2,000 turned out to vote 'For' (Table 6.1). Not sur-
prisingly, the proportion of respondents who voted 'For' decreases when
the bids get higher, with one exception: the proportion for the SEK 25,000
bid is somewhat greater than that for SEK 15,000, though not significantly
greater.[1] The fact that more than 10 per cent of the respondents that met the
highest bid voted 'For' implies that it is not clear what bid is required in or-

Table 6.1 *Responses to the WTP Questions*

Bids (SEK)	For	Against	Blank	Don't Know	Sum	No answer	Prop. For
0	345	4	9	29	387	23	0.891
1000	35	6	2	21	64	1	0.547
2000	29	6	1	18	54	1	0.537
4000	16	14	4	17	51	2	0.314
6000	15	20	4	18	57	0	0.263
10,000	11	19	7	23	60	0	0.183
15,000	6	23	8	14	51	2	0.118
25,000	10	30	4	22	66	1	0.152

1 The difference is not large enough to cause a rejection of the hypothesis of equal proportions.

der to cause a zero acceptance of the action plan. Another issue is whether more than 10 per cent of the respondents would in fact pay SEK 25,000 (about 12 per cent of their mean annual income) in reality. This cannot be taken for granted, which was also evident from telephone interviews with a few of the respondents who were willing to pay SEK 25,000. None of them was 100 per cent certain about their vote 'For', but on the other hand, nobody wanted to change their vote. On the whole, however, the use of conservative benefit estimates seems to be justified.

Why do people want to pay?

The respondents were asked to describe their motives for their willingness to pay for reduced eutrophication, see Söderqvist (1998) for an analysis of these motives. About one-third of the motives stated were related to direct use of the Baltic Sea, though often expressed in terms of future generations' (such as one's own children's use) rather than their own use. There was also a clear indication of the importance individuals place on the indirect use of ecosystems; almost 20 per cent of the motives stated indicated a perception of links between the health of the Sea and human health or survival. About 10 per cent of the motives can be described as a moral obligation to do something for the Baltic Sea. These motives may not be consistent with a teleological perspective, since they concern the action per se, rather than the consequences of the action. The other motives stated by the respondents include rather vague ones, for example, 'it's important' and 'it's important for the future' constituted about 18 per cent of all mentioned motives.

Protests against the valuation scenario

Some of the respondents objected to the valuation scenario. The analysis in Söderqvist (1998) showed that some protesters had more fundamental objections than others. The most common objections among these 'strong protesters' were along the lines of 'I'm innocent, the guilty ones should pay'. It is doubtful whether such respondents would accept having to make any type of payment for reduced eutrophication. There were also somewhat less fundamental objections that revealed scepticism about the success of the action plan and the reality of ear-marking funds for environmental improvement. Very few people raised objections to the idea of taking cost-effective actions, despite the fact that this implies that money collected in Sweden may be used for measures taken in other Baltic Sea countries. A group of 'weak protesters' was defined as the protesters that are likely to have a positive WTP for reduced eutrophication and accept a payment if the valuation scenario were to be changed slightly. Figure 6.1 illustrates the relative sizes of the groups of non-protesters (NP), strong protesters (SP), weak protesters (WP) and non-respondents (NR).

Estimation of national benefits

Söderqvist (1996a) employed different statistical methods in order to estimate mean WTP. The results indicated a mean WTP of non-protesters of

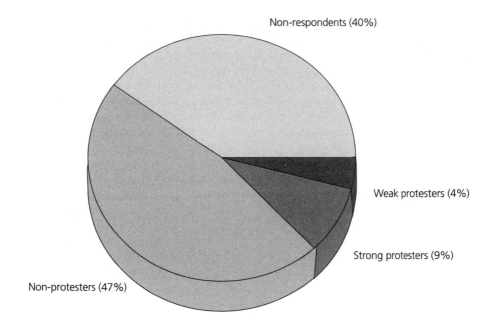

Figure 6.1 *Groups of Subjects in the Survey*

Table 6.2 *Present Values of the Conservative National Benefit Estimate*

WTP per adult Swede, year 1 (thousand SEK)	National benefits, year 1 (million SEK)	National benefits, present value[1] (million SEK)	National benefits, present value per year (million SEK)
3	19,893	210,750	10,537

[1] Time horizon: 20 years (as specified in the valuation scenario). Discount rate: 7. This rate was also used in the estimation of nutrient reduction costs in Gren et al (1995, 1997).

SEK 7000. Given an assumption of a zero WTP for all other subject groups, a conservative estimate of total national benefits of SEK 21,816 for year one is obtained. This corresponds to slightly more than SEK 3000 per adult Swede for year one. According to the valuation scenario, this payment would occur each year over 20 years. Given this time horizon and a discount rate of 7 per cent, Table 6.2 presents current values for these payments.

Other Swedish studies

The WTP question in the main CVM survey was framed as a DC question. An alternative is to let subjects state their WTP freely, ie, to use an open-ended (OE) WTP question. DC questions are presently favoured by most CVM practitioners and were also recommended by Arrow et al (1993). The far from undisputed arguments for DC questions include the resemblance to purchase decisions for ordinary market goods, and a low vulnerability to strategic behaviour. OE questions seem to result in lower benefit estimates (Kriström, 1993), but the reasons for this discrepancy are by no means well-established. The DC survey was therefore complemented by an experimental survey, identical to the DC survey except for the use of an OE question. The usual pattern concerning the size of benefit estimates was confirmed.

A TCM study was also carried out in Sweden, see Sandström (1996). It was based on a model of people's choice between recreational sites along the Swedish coast. Data on recreational behaviour were obtained from interviews with Swedes about their travel behaviour. A crucial issue is, however, whether eutrophication effects are one of the factors that influence the demand for recreation, and even if this is the case, what measure of eutrophication effects should be selected? It has to be an environmental quality index that is related to: (1) eutrophication as an objective phenomenon; and (2) something that people are actually able to perceive. Since increased turbidity is one important eutrophication effect, sight depth was selected as the quality index. This index turned out, in fact, to have a significant impact on travel behaviour in most of the estimated models. Moreover, a relationship between the sight depth and the concentration of nitrogen and phosphorus in water was confirmed by a statistical analysis.

The results of some policy simulations are also reported by Sandström (1996). One of the simulations was based on a 50 per cent uniform reduction of the nutrient load along the Swedish coastline. The estimated recreational benefits from such a change were, on average, about SEK 400 million per year, or about SEK 200 per visit. These are considerably lower benefit estimates than those obtained from the CVM survey. This is an expected result, as the recreational benefits estimated by the TCM study are only one of many possible types of benefits possible from reduced eutrophication. Moreover, the models used are likely to underestimate the recreational benefits. The fact that an improvement in water quality implies an increased value of each visit, may also result in an increase in the total number of visits. The latter effect is not taken into account by the estimated models. The magnitude of the difference between the CVM and TCM estimates is, however, large enough to motivate further analysis.

Lithuania and Poland

A number of CVM Baltic Sea surveys have been carried out in Poland, and one survey was also implemented in Lithuania, see Chapter 7 for details. These pioneering CVM studies for formerly centrally-planned economies employed various valuation scenarios, elicitation methods and survey methods. The focus here will be on the Polish mail questionnaire survey because of its close resemblance to the Swedish survey. The valuation scenario was almost identical. It used a DC elicitation format and the design of the questionnaire was also, in most other respects, parallel to the Swedish one.

The Polish mail CVM questionnaire was sent out to 600 randomly selected Polish adults. Completed questionnaires were received from slightly more than 50 per cent of the sampled individuals. After an examination of the reasons for refusals to the idea of paying a tax for the Baltic Sea, and a logit analysis of answers to the WTP question, Markowska and Zylicz (1996) report a mean annual WTP of SEK 600 per person when subjects who protested against the valuation scenario are assigned a zero WTP. In order to obtain a conservative estimate comparable to the Swedish one, a zero mean WTP of non-respondents is assumed. This results in a mean WTP estimate of about SEK 300 per person per year.

BASIN-WIDE BENEFITS

The Polish and Swedish CVM results will now be used for an estimation of basin-wide benefits. Needless to say, this is an exercise that relies on very strong assumptions. The aggregate benefit estimates to be presented below should thus not be taken too literally. However, they may give useful information about the order of magnitude of the basin-wide benefit estimates. Given an adjustment for the difference in GDP per capita levels between the countries, the Polish mean WTP estimate of SEK 300 will be regarded as representative of the transition economies around the Baltic Sea – Estonia, Latvia, Lithuania, Poland and the Russian Federation – and the Swedish mean WTP estimate of SEK 3000 as representative of the established market economies – Denmark, Finland, Germany and Sweden. The possible WTP of

Table 6.3 *Basin-wide Benefit Estimates*

Country	Mean annual WTP per person[1] (SEK k)	National benefits, year I[2] (SEK m)	National benefits, present value[3] (SEK m)	National benefits, present value per year (SEK m)
Transition economies:				
Estonia	0.375	423	4476	224
Latvia	0.257	497	5260	263
Lithuania	0.182	477	5050	253
Poland	0.3	7842	83,077	4154
Russian Federation	0.246	1782	18,883	944
Market economies:				
Denmark	3.515	12,131	128,514	6426
Finland	2.229	8369	88,661	4433
Germany	3.334	8104	85,852	4293
Sweden	3	19,893	210,750	10,537
Total		**59,518**	**630,523**	**31,527**

[1] For the transition economies, the Polish mean WTP estimate of SEK k 0.3 was multiplied by the ratio between each country's GDP per capita (at purchasing power parity (PPP)) and Poland's GDP per capita at PPP. For the market economies, the Swedish mean WTP estimate of SEK k 3 and Sweden's GDP per capita at PPP were used correspondingly. The GDP per capita at PPP estimates employed correspond to those used by Markowska and Zylicz (1996, p. 25).
[2] The mean annual WTP estimates per person multiplied by the (adult) population in the Baltic Sea drainage basin part of each country. Population data were obtained from Sweitzer et al (1996).
[3] Time horizon: 20 years (as specified in the CVM studies). Discount rate: 7%. This rate was also used in the estimation of nutrient-reduction costs in Gren et al (1995, 1997).

the population in the other countries included in the Baltic Sea drainage basin (Belarus, the Czech Republic, Norway, Slovakia and Ukraine) is disregarded here.

In order to calculate national benefit estimates, the estimates of mean WTP per person were multiplied by the (adult) population in the Baltic Sea drainage basin area of each country. In the case of Germany at least, this is likely to cause a considerable underestimation, as the German users of the Baltic Sea also live in other parts of Germany. According to the results presented in Table 6.3, the basin-wide estimate for total benefits from a reduced eutrophication of the Baltic Sea amounts to slightly more than SEK 31,000 million per year.

CONCLUSIONS

The benefit estimation exercises described in this chapter were, in many senses, a speculative venture. In particular, this is true for the CVM studies. To communicate adequate and sufficient information on a complex phenomenon such as the eutrophication of the Baltic Sea and to design a reasonably realistic valuation scenario involving a reduction of the eutrophication are far from trivial tasks. Moreover, to take a stand on the value of the benefits of an action plan to combat eutrophication is not an everyday situation for citizens, and a lack of familiarity is often claimed to impair the working of the CVM. These circumstances call for cautious interpretations of the results of the study.

One way to be cautious is to compute conservative benefit estimates, and this route was followed in this chapter. It should be remembered that these benefits are perceived benefits, and are not necessarily equal to the total benefits from a less eutrophicated Baltic Sea. The estimated benefits are nevertheless substantial, and very close to the estimated total costs for

Table 6.4 *Benefits and Costs from a 50 Per Cent Reduction of the Nutrient Load to the Baltic Sea, Millions of SEK Per Year*

Country	Reduction in %	Benefits	Costs	Net benefits
Transition economies:				
Estonia	55	224	1529	−1305
Latvia	56	263	1799	−1536
Lithuania	55	253	2446	−2193
Poland	63	4154	9600	−5446
Russian Federation	44	944	586	358
Market economies:	51	6426	2962	3464
Denmark	52	4433	2838	1595
Finland	39	4293	4010	283
Germany	42	10,537	5300	5237
Sweden				
Total	**50**	**31,527**	**31,070**	**457**

an international action plan involving a cost-effective 50 per cent reduction of the total nutrient load to the Sea (Table 6.4). The cost estimates in this table also include the costs of reducing the total load by 50 per cent from all countries, and not only the load from the countries located in the drainage basin as shown in Chapter 4. A 50 per cent reduction in the total load corresponds to almost a 60 per cent reduction in the nutrient load from the countries located in the drainage basin, see Figure 4.1 in Chapter 4. The national net benefits for the established market economies (including Sweden) seem to be positive, whereas the opposite is true for most of the transition economies. On the whole, the total net benefits seem to be positive. These results show, however, smaller benefits than those presented in Chapter 7.

While it is true that the valuation scenarios used in the CVM studies are not completely different from policies actually discussed for the Baltic Sea region, we are still dealing with an extremely simplified situation. There seems to exist a substantial willingness among Poles and Swedes to make sacrifices (or should it be called investments?) for the sake of the Baltic Sea. However, it is an inescapable fact that this observation is made in an institutional vacuum. How can institutions be built that allow the realization of a large-scale and cost-effective action plan? The comparison of costs and benefits in Table 6.4 suggests that side payments between countries are necessary for accomplishing cost-effectiveness. Will it be possible to design such a payment scheme? And how is a cost-effective action plan to be implemented? These are issues that are analysed in the following three chapters.

Part III

INSTITUTIONS AND POLICIES

7 Policy Instruments and Cost Sharing of Baltic Sea Cleaning Programmes

Agnieszka Markowska, Tomasz Zylicz and Ing-Marie Gren

According to our data and analysis, it is reasonable to conclude that, in the context of the ministerial agreement on a 50 per cent nitrogen reduction, Baltic peoples' willingness to pay almost matches the necessary costs. Thus, applying the economic benefit-cost test, the project should be undertaken. However, the complex question of how to implement the project remains to be answered. In this chapter, two questions are raised: 1) what would be the charge levels, or prices of permits under a system with tradable permits, in order to implement the cost effective programme, and 2) if one or several countries will not make payments to cover the costs of cleaning measures without receiving any compensation, how should an appropriate cost-sharing scheme be designed.

In principle, we therefore have two types of economic instruments capable of generating the correct incentives for implementing the cost-effective programme suggested in Chapter 4; charges and permit markets. Simple applications of these policy schemes will be discussed in this chapter under the assumption that the countries are willing to make the cost-effective allocation of payments. This assumption is relaxed when the properties of an efficient cost-sharing programme are investigated. The cost-sharing programme requires information on both costs and benefits, which were presented in Chapter 6. In addition to policy schemes, this chapter provides alternative benefits estimates based on specific Polish valuation studies.

Cost-effective Charges

Under a charge system, each firm reacts by undertaking cleaning activities up to the level where the marginal cost equals the charge level. This simple idea is illustrated in Figure 7.1, where MC shows the marginal cost curve. That is, the change in total costs of 1 kg nitrogen reduction. Usually this curve is increasing, since a cost-minimizing firm starts by adopting the cheapest cleaning technology, and then the next cheapest, and so on.

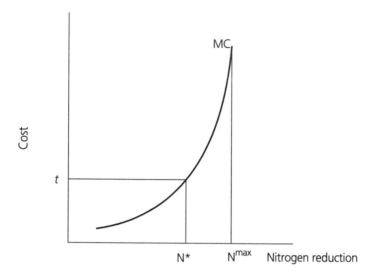

Key: MC = marginal cost curve
       N* = nitrogen reduction level at the charge *t*
       N^{max} = maximum nitrogen reduction level
       t = charge per kg nitrogen

Figure 7.1 *Charge and Nitrogen Reduction*

If the firm faced no costs because of its nitrogen emission, there would be no nitrogen emission reductions. The firm would then emit N^{max}, which also corresponds to the maximum amount of nitrogen that can be removed by the firm. Assume now that the charge *t* per kg nitrogen reduction is implemented. The firm then responds by reducing the nitrogen emission to the level *N**. The reason is that the firm is assumed to always chose the cheapest option. The firm reduces nitrogen emission as long as the charge is higher than the unit cost of a marginal nitrogen reduction, ie to the left of *N** in the figure. When the charge is lower than the unit cleaning cost, the firm choses to emit nitrogen, which, in the figure, occurs at the right of *N**. Thus, as long as there is a difference in the charge and the unit cost of a marginal nitrogen reduction, the firm makes net gains by either increasing or decreasing nitrogen reduction. Only when *t=MC* does the firm have no incentive to change the nutrient-reduction level. Since the charge is the same for all firms, it is also the case that they reduce their nutrients to the level where *MC* is equal for all firms. From Chapter 4, we know the level at which the total nutrient reduction is cost-effective.

The simple illustration in Figure 7.1 shows that once we have information on the marginal costs at the 50 per cent reduction levels of nitrogen and phosphorus respectively, we are able to identify the cost-effective charge.

However, this charge measures the marginal cost of reducing nutrient loads to the Baltic Sea. But, as demonstrated in Chapter 2, the emission sources differ with respect to the transport of nutrient loads to the coastal waters. So called point sources, such as sewage treatment plants and industry, are located along the coasts. Their nutrient effluents are discharged directly into the coastal water. Other emission sources, non-point sources, are located upstream and their discharges into the coastal water follow much more complex pathways via surface and groundwater water in the drainage basin. This also means that only a fraction of their nutrient emissions finally reaches the coastal waters. The recognition of these differences between different nutrient sources requires an adjustment of the charges in order to account for the differences in transport processes to the coastal waters. The general rule is to charge emission sources in terms of their effect on the target. For example, assume that only one-half of the nitrogen emissions from a distant non-point source reaches the coast. Then, at the cost-effective charge of SEK 100/kg nitrogen emission, this source should face the effective charge of SEK 50/kg nitrogen emission. The charges on point-source emission is the same for all countries, as the impact of this source is the same, assuming all the nutrient effluent is discharged directly into corresponding coastal waters. However, the effective charges on non-point emission sources differ between regions due to differences in impacts per unit of nutrient emissions, which are reported in Table 2.2. The effective charges for the non-point sources can thus be calculated by means of Table 2.2 and the estimated marginal costs of SEK 149/kg nitrogen emission and SEK 174/kg phosphorus emission at the cost-effective 50 per cent nitrogen and phosphorus reductions respectively. These effective charges faced by the non-point emission sources are presented in Table 7.1.

Table 7.1 *Effective Charges at Non-point Nutrient Emission Sources, SEK/Nitrogen and Phosphorous Emission Respectively, at 50 Per Cent Nutrient Reduction Targets*

Regions	Nitrogen emissions:			Phosphorous
	Rivers	NO_x	NH_4	emissions
Finland:				
Bothnian Bay	16.4	4.5	4.5	19.1
Bothnian Sea	16.4	4.5	4.5	19.1
Gulf of Finland	14.9	4.5	4.5	17.4
Sweden:				
Bothnian Bay	6.0	19.4	19.4	7.0
Bothnian Sea	10.4	20.9	19.4	12.2
Baltic Proper	20.9	23.8	23.8	24.4
Russian Federation				
Kaliningrad	7.5	4.5	2.9	8.7
St Petersburg	6.0	4.5	2.9	7.0
Poland	7.5	8.9	7.5	8.7
Lithuania	2.9	8.9	7.5	3.5
Latvia	4.5	13.4	11.9	5.2
Estonia	4.5	14.9	17.9	5.2
Denmark	14.9	22.4	20.9	17.4
Germany	8.9	8.9	6.0	10.4

From Table 7.1 we can see that the effective charge per kg nitrogen emission ranges between 2.9 and 23.8, with the highest charge relating to non-point sources in the Swedish drainage basin discharging to the Baltic Proper. The relatively low nitrogen emission charges are found in regions with high rates of nutrient retention, such as the Russian Federation, Lithuania, Latvia and Estonia. The range of phosphorus emission charges is about the same as for the nitrogen charges, between SEK 3.5 and 24.4, still being highest for Baltic Proper and lowest for Latvia and Estonia.

BALTIC SEA NITROGEN AND PHOSPHORUS PERMIT MARKETS

The basic idea of a permit market system is to let the market for permits determine the price of nitrogen and phosphorus load permits. Pollution control authorities then only have to regulate the total amount of nitrogen and phosphorus loading. In our case, this corresponds to 50 per cent of these loads in 1994. Each polluter receives or purchases initial permits. In principle, if the market functions as it should – meaning no single buyer of permits or group of buyers can influence the price of permits – there is no difference between the two systems for initially distributing permits in terms of permit allocation and prices (Montgomery, 1972). However, there is a difference with respect to incomes for the governments, who receive revenue when the initial permits are sold, but not otherwise.

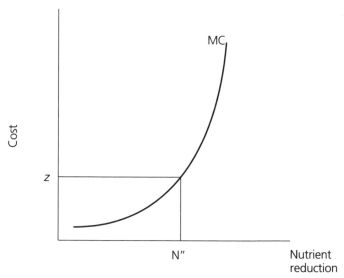

Key: MC = marginal cost curve

  N″ = required reduction in nutrient

  z = price of permits per unit of nutrient reduction

Figure 7.2 *Nutrient Permit Market*

Assuming that the permit market functions as it should, the price of permits for nitrogen or phosphorus emissions is determined as illustrated in Figure 7.2, where $N"$ denotes the required reduction in nutrient loads to the Baltic Sea, and MC is the marginal cost curve for nutrient load reductions.

The MC curve can be regarded as reflecting the demand for nutrient emission permits. The reason is that when the price of permits per unit nutrient reduction, z, is higher than MC it is more profitable to undertake nutrient reductions and sell permits. When the marginal cost is higher than the permit price, it is better to meet the requirement for reduction by using permits sold by somebody else. An equilibrating permit price then is equivalent to the marginal cost at the required reduction $N"$. To the left of $N"$, we have $MC<z$ which implies a surplus supply of permits, since it is more profitable to clean up than to use permits. The opposite is the case to the right of $N"$, where permits are used for nutrient emissions. Similar to a charge system, we also have cost-effective $N"$ since the permit price z reflects the marginal nutrient reduction cost of all firms.

Permit markets were brought up in public debate and actually implemented in the US in the mid 1970s (see Tietenberg, 1995, for a review). Over a period of about 25 years, several practical problems have arisen, such as how to deal with market power and ambient environmental standards. If some firms involved can influence the price of permits, the permit price will be higher than z, implying that the pollutant reduction is not cost-effective. Another difficulty relates to the fact that the location of firms influences the environmental impact generated by their pollutant emissions. In order to achieve certain ambient standards, firms would have to face different permit prices depending on their location and related environmental impact. One way to avoid such difficulties is to trade permits according to specific exchange rates. Then, instead of establishing a permit market where permit prices are determined, firms would be allowed to trade permits among themselves. The trading ratio between different sources can then reflect their environmental impacts. For example, if a reduction in nitrogen from a non-point source by 1 kg nitrogen implies 0.5 kg nitrogen reduction into the coastal water, the trading ratio between this non-point source and a coastal sewage treatment plant which discharges nitrogen directly into the water corresponds to 2:1. That is, if the sewage treatment plant considers trading with the non-point source, it must have at least 2 permits in exchange for 1.

As the drainage basins of the Baltic Sea differ with respect to transport of nutrients, the impacts of non-point sources also differ in terms of their transport to the coastal waters. Given that our target is as formulated in Chapter 4 – the required nutrient reductions in the coastal waters – the impacts of the point sources are the same in all countries and regions. Trading ratios then need to be defined only for non-point sources in each drainage sub-basin, which can be calculated from the impacts data presented in Table 2.2. Such trading ratios are shown in Table 7.2.

The trading ratios between non-point/point source emissions in Table 7.2 reveal the quite large differences in impacts between non-point sources in the drainage basins. For example, the impact of non-point source nitrogen emissions in Denmark is seven times as high as that of a corresponding

Table 7.2 *Trading Ratios for Non-point/Point Nitrogen and Phosphorus Emission Rights*

| Region | Nitrogen emissions: | | | Phosphorous emissions |
	Rivers	NO_x	NH_4	
Finland:				
Bothnian Bay	8.9	33	33	9.1
Bothnian Sea	8.9	33	33	9.1
Gulf of Finland	9.9	33	33	10
Sweden:				
Bothnian Bay	24.5	7.6	7.6	24.9
Bothnian Sea	14.1	7.0	7.6	14.3
Baltic Proper	7.0	6.2	6.2	7.1
Russian Federation				
Kaliningrad	19.6	32.7	50.7	20
St Petersburg	24.5	32.7	50.7	24.9
Poland	19.6	16.5	19.6	20
Lithuania	18.6	16.5	19.6	49.7
Latvia	32.7	11	12.4	33.5
Estonia	32.7	9.9	8.2	33.5
Denmark	9.9	6.6	7.0	10
Germany	16.5	16.5	24.5	16.7

source in the Russian Federation. If trade between non-point sources among the countries is considered, the results in Table 7.2 can be used to calculate associated trading ratios. In the case of Denmark and the Russian Federation, one nitrogen-emission permit in Denmark corresponds to approximately seven permits for non-point sources in the Russian Federation (which is calculated as 50.6:7).

AN EFFICIENT COST-SHARING SCHEME

Hitherto, it has been assumed that the countries are willing to carry out the cost-effective nutrient reductions without any compensation. This assumption is, however, questionable in the light of past experience. By the very end of the second decade of the Helsinki Convention, in April 1992, its signatories committed themselves to the Joint Comprehensive Environmental Action Programme (JCEAP). The Programme lists 132 hot spots in the drainage basin (Helcom, 1993, no. 46) and indicates priority investment projects necessary to bring about an improvement. At the same time, however, it failed to create an effective mechanism for these projects to be undertaken. In some cases, these projects have already been launched, but some clearly need an additional stimulus before they will actually materialize. It has been recognized that countries need to create a basin-wide compensation mechanism to implement the JCEAP. Indeed, since 1990, some transfers between the wealthy north-western and less economically developed south-eastern parts of the drainage basin have taken place. Nevertheless, it is far from certain whether the absolute amount of these transfers and their breakdown by donors and recipients, is consistent with any conscious policy to optimize the Baltic clean-up process.

An efficient international cost-sharing scheme has been suggested by Chander and Tulkens (1992). The idea underlying the model is to theoretically develop an international financing mechanism that would be both cost-effective and equitable for all the countries that share responsibility for the public good, and that are concerned about its quality. The cost-effectiveness criterion, as defined in Chapter 4, requires that the countries with lower pollution abatement costs should abate more, while the countries with relatively high marginal costs of abatement (due mainly to the fact that all the cheaper options have already been exploited) should abate relatively less. This ensures that the overall cost of achieving a certain abatement goal will be minimized. This scheme may, however, be changed – weakened or strengthened – after taking into consideration the benefits side of the model. The second criterion, international equity within the region, may be satisfied by looking at the relative benefits accruing to each country, that result from utilizing the common good. In principle, if a given country receives very high benefits from utilizing the common good, it should contribute proportionately more to the pollution abatement effort than another country that does not gain as much benefit. This contribution may take the form of co-financing abatement in countries where costs are relatively low.

More specifically, within the framework of the Chander and Tulkens model, we could write a general, simplified formula describing the amount of money transfers for a country, T^i, as follows:

$$T^i = C^i - (\pi^i/\pi^n) \times C^n$$

where C^i is the cost incurred by country i, π^i is the marginal benefits of country i, π^n is the sum of marginal benefits for all countries and C^n is the sum of marginal costs of all the countries. According to the scheme, the countries have fully covered their respective costs of abatement (C^i) according to the cost-effective scheme (see Chapter 4) but at the same time they give away:

$$(\pi^i/\pi^n) \times C^n$$

ie a share in the sum of all countries' additional costs (C^n) proportional to their share in the sum of marginal benefits of all countries.

The approach taken in this chapter is as follows. We make a simplifying assumption that the Baltic Sea eutrophication problem is caused by nitrogen emissions only. The cost-sharing scheme for nitrogen reductions is then based on the cost estimates reported in Chapters 4 and 6. The benefits estimates used are based on Polish valuation studies, which are briefly reviewed before the calculations of transfers between countries are presented. In central and eastern European countries, methods of economic valuation based on demand-side data were not utilized until the beginning of the 1990s. The interest in using consumer-surplus theory as a valuation tool has emerged together with the embryonic eastern European market economies. A series of contingent valuation surveys carried out under the Baltic Drainage Basin Project (funded by the European Commission) are the first examples of a full-blown CVM study in such a context.

The main Polish study was carried out by a professional polling agency on 25–29 November 1994, using a representative all-Poland sample of 1162 respondents, as part of a larger sociological survey. A face-to-face interview method was used. A valuation scenario gave a short description of the negative eutrophication effects observed in the Baltic Sea and mentioned the possibility of carrying out an international clean-up action over the next ten years. Such an action would require financial resources that could be collected in every Baltic country in the form of a special tax. The respondents were asked if they would support such an action. Some 62.5 per cent of respondents said 'yes' (26 per cent 'definitely yes' and 36.5 per cent 'probably yes'); and 29.8 per cent said 'no' (19.8 per cent 'possibly not' and 10.9 per cent 'certainly not'); 6.7 per cent said they did not know.

A referendum question asking for acceptance/rejection of an initial bid followed. Initial bids (randomly distributed across the sample) and the ratio of acceptance/rejection grouped according to initial bids are displayed in Figure 7.3. The numbers displayed in the bars represent percentage of yes/no answers for groups of respondents that were given a particular initial bid.

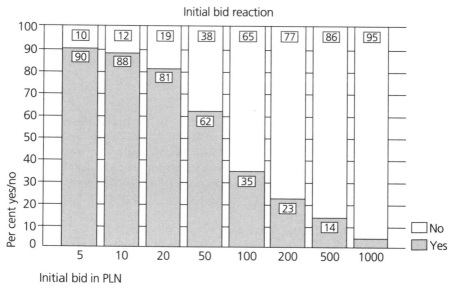

Figure 7.3 *Initial Bids and Acceptance Rates*

As expected, the acceptance rate falls as initial bids rise. For the estimation of mean and median values of WTP, a so-called linear logit model was used with one explanatory variable, bid level.

Based on the model co-efficients we estimated, the mean value at 169.33 PLN (≈SEK 546). However, this value may only be relevant for the group of respondents who answered the question about their support for the Baltic tax in the positive. In order to make the distinction between those negative respondents who presumably expressed zero value and others, who were reluctant to pay for some other reasons, we have investigated the reasons for rejecting the 'Baltic tax' concept. Table 7.3 provides a summary of the reasons for refusal, with percentage of all refusals, as well as the breakdown for 'positive', 'protest', and 'zero-bidders' as set out in Bergland et al (1995).

Table 7.3 *Refusal Rationales*

• Zero bidders (those with presumably low or nil WTP)	

I cannot afford to pay but would do so otherwise 45.4% of all refusals

There are other, more important needs 5.2% of all refusals

• Protest bidders (those with presumably positive WTP, but who did not accept the survey)
I pay too much taxes already 16.5%
I am not responsible for the state of the Baltic Sea; let those responsible pay 14%
I do not believe that my contribution will lead to any improvement 13.8%
Other reasons 3.6%
Difficult to say 1.6%

	Frequencies	% of the whole sample
zero bidders	224	19.3
positive bidders	713	61.3
protest bidders	225	19.3
Total	**1162**	**100**

In our calculations, protest bidders are treated as missing values and zero bidders as zero WTP. The mean value for the sample including zero bidders has been calculated at 129 PLN (≈SEK 416).

Table 7.4 shows WTP values obtained from the various studies. Altogether, three studies from Poland are included (a pilot study, a main study as described above, and a mail study; one study from Lithuania (very similar to the Polish pilot study); and one from Sweden (using the mail method). The studies employed different questioning methods, OE indicates an open-ended question about WTP, while DC stands for dichotomous choice (yes/no referendum) question.

We stress that these values should not be treated as precise estimations of the value of the Baltic clean-up. At the same time, they are the only estimates available to approximate the (π^i/π^n) ratio to be used in the Chander-Tulkens model.

As we had the results from studies that employed three different methods of estimation, we have translated the results from the Lithuanian OE study and Swedish mail DC study into DC face-to-face type of study results, using the Polish experience from all three types of surveys. Thus, based on

Table 7.4 *CVM Studies Results – Mean WTP Values in SEK*

Country	Lithuania		Poland		Sweden
Type of empirical study	Pilot	Pilot	Main	Mail	Mail
	OE	OE	DC	DC	DC
positive bidders only	67	127	547	1552	5797
zero and positive bidders	52	105	416	765	3439
whole sample (0 for refusals)	37	75	337	600	3600

Table 7.4 above, we have estimated that the co-efficient for translating an OE study result into a DC face-to-face study result equals four, and for translating mail survey results the co-efficient is 0.55. Finally, we assume that within each of the following groups of countries (sub-regions) WTP is proportional to GDP per capita (at PPP). Six extrapolations have thus been made. The Swedish WTP was extrapolated to Finland, Denmark and Germany using ratios of the Swedish GDP and those of respective countries. Likewise, the Lithuanian WTP was extrapolated to Latvia, Estonia and the Russian Federation using ratios of the Lithuanian GDP and those of respective countries. The GDP figures and their ratios (β_i) to the GDP of the reference country for each group, are listed in Table 7.5 below. Table 7.5 contains WTP per capita estimates based on the WTP attributed to a reference country and modified by the β_i co-efficient. Aggregate WTP for countries, π_i, are then calculated by multiplying mean WTP per capita by the number of adults living in respective countries' Baltic drainage basin (Sweitzer et al, 1995; Statistical Yearbook, 1995).

The calibration of the Chander-Tulkens formula is given in Table 7.6 below. The cost estimates applied are presented in Chapter 6, Table 6.4. According to the model, there is no country within the region that achieves balance without a money transfer with nutrient load reductions as shown in

Table 7.5 *Estimation of Mean and Total WTP Values in Countries of the Baltic Region*

Country	Nominal GDP per capita (SEK)	GDP per capita at PPP (SEK)	β_i	Mean WTP per capita (SEK)	π_i (10^6 SEK)
Finland	104,655	116,122	0.92	1740	6540
Sweden	133,327	126,157	1	1890	12,113
Denmark	163,432	144,795	1.15	2168	7478
Germany	147,660	130,058	1.10	2085	5070
Poland	14,332	34,410	1	420	10,950
Lithuania	4297	27,240	1	210	547
Latvia	5737	22,935	0.84	180	345
Estonia	7170	28,672	1.05	217	247
Russian Federation	8603	37,275	1.37	285	2070
Drainage Basin Region	**45,682**	**59,985**	–	**825**	**45,360**

Source of GDP data: OECD

Table 7.6. In four cases (Finland, Sweden, Denmark and the Russian Federation) countries should contribute to a 'common pool' to be devoted soley to the Baltic clean-up (negative T^i values), and in five cases (Germany, Poland, Lithuania, Latvia and Estonia) countries should be able to make withdrawals from the 'common pool'.

Table 7.6 *Hypothetical Money Transfers T'_i in the Chander-Tulkens Model*

Country	π_i / π_N	T'_i $(10^6$ SEK)
Finland	0.144	−1627
Sweden	0.267	−2967
Denmark	0.165	−2192
Germany	0.112	504
Poland	0.241	2106
Lithuania	0.012	2100
Latvia	0.008	1566
Estonia	0.006	1329
Russian Federation	0.046	−819
Total	**1.000**	**0.00**

The total amount of transfers, ie $\frac{1}{2}\Sigma_i T^i$ (for each crown transferred there is one paying and one receiving party, but we count the amount only once), equals SEK 7605 million per year which corresponds to almost 23 per cent of the costs. On the other hand, these transfers correspond to 16 per cent of the annual benefits. As one would expect, north-western countries such as Sweden, Denmark and Finland are net payers owing to the fact that they value the Sea more than their share in abatement expenditures. By the same token, south-eastern countries such as Poland, Lithuania, Latvia and Estonia experience net gain. Quite unexpectedly, two countries, Germany and Russia, end up in 'wrong' groups: the Russian Federation is a net payer (SEK 819 million, ie by 30 per cent more than it spends on abatement, or 40 per cent of its benefits) while Germany receives a small compensation (SEK 504 million, ie 13 per cent of its abatement costs or 10 per cent of its benefits). The German-Russian Federation paradox can be explained by taking a closer look at our estimates π^i of these countries' WTP. Only a very small part of the German population lives in the Baltic drainage basin and was therefore not considered in the benefits study. For instance, the population of Berlin, living just a two-hour drive from the Baltic coast, has been excluded from the benefit-aggregation procedure. And yet, if asked, they would probably have revealed a substantial WTP for improving the marine environment. It is therefore likely that the German benefits are greatly underestimated in our study.

At the same time, the Russian WTP was estimated by taking into consideration the country's GDP per capita at PPP. There is an unusual discrepancy between this estimate (almost SEK 37,500) and the nominal GDP (SEK 8602). Hence, one may expect that the former measure is not a very accurate one of the wealth of an average Russian. Moreover, it is likely that he or she would be inclined to relate WTP to an international public good such as the Baltic Sea, by reference to nominally priced private goods rather than by considering his or her purchasing power in real terms. As a result, the Russian Federation WTP estimated in our study could have been overestimated. Another way of looking at the German-Russian Federation paradox is to observe that both countries are allocated lower than average reduction targets (39 per cent and 44 per cent, respectively, instead of the uniform 50 per cent). In this way, they may save on abatement costs but, they may have contributed to financing other countries' expenditures. The German obligation is reduced and even reversed by the very low, perhaps underestimated, WTP attributed to this country. In the Russian Federation case no such countervailing tendency exists.

DISCUSSION AND POLICY CONCLUSIONS

In this chapter, simple Baltic Sea cost-effective charge and permit market systems were investigated. It turned out that charges and permit values are quite different across the drainage basins due to their differences in terms of nutrient transports. The effective charges for non-point sources may vary between approximately SEK 3 per kg and 24 per kg of nitrogen emission. The corresponding variation of the effective phosphorus emission charge ranges between SEK 5 per kg and SEK 24 per kg. A simple investigation of a potential permit market also reveals relatively large differences between the 'values' of permits, depending on where the permits is used. For example, a permit for non-point source emission in Denmark corresponds to seven non-point source permits in the Russian Federation.

However, it is quite likely that an international agreement on any policy scheme will require some kind of money transfers between the countries, in order to be accepted. This chapter suggested one such scheme, the Chander-Tulkens model. While a large margin of uncertainty surrounds the benefit estimates, the series of surveys aimed at the valuation of the same public good still revealed a fair degree of internal consistency and reliability. All surveys carried out were based on the same valuation scenario – to reduce the eutrophication of the Baltic Sea to a tolerable level, defined in terms of reduced algae bloom episodes, eliminated oxygen deficit conditions, and re-opened beaches. All surveys referred to the same payment vehicle – an internationally-coordinated, ear-marked Baltic tax.

As anticipated by the economic theory, we found that OE questions yielded consistently lower WTP responses. At the same time, we found that mail surveys yielded higher WTP responses than face-to-face ones. We finally decided to base our conclusions on a face-to-face survey with a referendum WTP question for two reasons. First, this type of survey produced mid-point results. Second, it seemed to provide information sufficient to define the commodity to be valued without leaving the respondent overwhelmed by the message.

Our benefit-transfer exercises demonstrated that the WTP revealed in various countries was strongly correlated with the GDP per capita. However, they also revealed that monetary wealth alone is not sufficient to predict inter-country differences. The econometric model, whose parameters were estimated on a combined Polish-Lithuanian sample, proved that neither the income nor other socio-economic characteristics of respondents were sufficient to explain the variance in WTP responses. A dummy variable to indicate respondents' home country turned out to be highly significant. Consequently, we decided to minimize the need for benefit transfers by choosing three countries rather than one, as extrapolation bases. We chose Swedish WTP regression equations as representative of the other north-western countries – Finland, Denmark and Germany. We chose Lithuanian results as representative of the countries of the former Soviet Union – Latvia, Estonia and Russia. We have not extrapolated Polish data to other countries. Poland's population accounts for 45 per cent of the total drainage basin, and 19 per cent of the area studied in this paper (without the 'Baltic' parts of Norway, the Czech Republic, Slovakia, Ukraine and Belarus).

Sweden emerges as the single largest beneficiary of the clean-up programme, with an almost 30 per cent share in total benefits. This is followed by Poland (almost 24 per cent), and Denmark (16 per cent). The Baltic States are located at the bottom of this list, enjoying collectively less than 3 per cent of the total benefits. They are marginally surpassed by Russia and by Finland and Germany by a factor of five and four respectively. Thus more than 6 per cent of all benefits accrue to Sweden, Denmark and Poland, either because of the high average WTP (Russia and Finland), or because of the large population affected (Germany). It therefore comes as no surprise that these three countries have been the most active in promoting a coordinated, basin-wide approach to the problem at various Helsinki Commission negotiations.

Another important observation emerging from our analysis is that the total benefits π^n from clean-up (SEK 45,360 million per year) outweigh the costs C^n (SEK 30,843 million per year) by almost a factor of 1.5. A similar pattern was found by Gren et al (1995) although in that study the benefit to cost ratio was estimated at 1:13 only. The main difference between these two assessments has been caused by our method of estimating mean WTP in the economies in transition. In the Swedish study, the Polish WTP was estimated at 41 per cent of the Swedish one and it was then extrapolated proportionally to other economies in transition in relation to their respective nominal GDP per capita values. Having scrutinized extensive evidence available from a series of Polish studies, we concluded that the mean Polish WTP corresponds to 22 per cent of the Swedish one. Incidentally, this is higher than the ratio of nominal GDP per capita indicators (11 per cent), but lower than the same ratio estimated at PPP (27 per cent). Cost estimates taken from Gren et al (1995) refer to average levels annualized over a 20-year period. They are higher than costs to be borne in the first few years of the JCEAP, but lower than those to be borne towards the end of the programme.

A rough calibration of the Chander-Tulkens model allowed us to compute hypothetical money transfers required to motivate countries to participate in the JCEAP up to the globally-efficient level. Four countries are the net payers, while the other five gain in net terms. In other words, under a concerted clean-up programme, the former group of countries, on the margin, enjoy a larger share of the benefits from reduced eutrophication of the Baltic Sea than they pay in costs for abatement measures. On the other hand, the latter group abates more than it benefits from the improvement in the quality of the Sea as a common international good. Money transfers are introduced to restore the balance.

Only fragmentary evidence of the actual environment-related money transfers is available (Laurson et al, 1995). Since 1989, Poland has attracted more international environmental assistance than other countries – approximately SEK 300 million to SEK 375 million per year. In relative terms, however, this accounts for a mere 4–5 per cent of the country's environmental investment expenditures. Interestingly for Poland, Denmark turns out to be the largest single European donor, with grants pledged totalling SEK 265 million over a five-year period (1990–1994). Sweden pledged SEK 165 million and Finland SEK 55 million. Germany considered pledging SEK 213 million, but only SEK 0.75 million was actually spent by 1994. The Polish case is rather typical for the region, with the exception of Estonia. The latter has received substantial Finnish assistance totalling almost one-half of its environmental expenditures in the early 1990s. At least a part of this assistance can be attributed to the Baltic preferences revealed by Finland, but there are no other examples of that sort and of that order of magnitude. Sweden, the single most generous international donor (in per capita terms) with a strong interest in the region, committed SEK 1050 million per annum (1993–95) to projects in central and eastern Europe. However, only a minor fraction of these funds was to be spent on the environment in competition with a number of other (political, educational, cultural, medical and commercial) objectives (Ministry of Finance, 1992).

It is absolutely clear from these examples that actual transfers between the Baltic drainage-basin countries are at least lower by one order of magnitude than those derived from the Chander-Tulkens model. Nevertheless, a conclusion that countries do not work towards an efficient level of the provision of the public good would not be warranted for two reasons. First of all, the countries of the region are involved in a number of economic, social and political issues that are far more complicated than the Baltic question. It can thus happen that 'issue linkage' occurs and other forms of incentives or pressures substitute for direct transfers. In addition, cost estimates that were taken as the basis for further modelling represent an approximate average level required until the year 2010 (the time horizon of the JCEAP). Given the steepness of the underlying abatement cost curves, the largest effort has yet to be undertaken. So far, countries have not left the relatively flat portions of their marginal cost curves. Hence, it is likely that the actual level of transfers in the early 1990s was not very far from the hypothetically-efficient one.

The latter argument, however, leads to the conclusion that, if in the next decade, countries are to efficiently allocate their abatement effort, they must be ready to establish a financial mechanism to carry out transfers on a much larger scale than currently. According to the model predictions, an efficient level of transfers corresponds to more than 22 per cent of the total region-wide costs. In the mid 1990s there were proposals raised at the Helsinki Convention for the establishment of a Baltic-wide fund to stimulate an improved allocation of abatement effort. Even though the proposals were very modest ones (corresponding to a mere 5 per cent of abatement costs) they have not gained much political support thus far. This casts serious doubt on whether the signatories of the Helsinki Convention are ready to adequately provide and pay for their Baltic Sea international public good.

8 WINNERS AND LOSERS FROM BALTIC SEA NITROGEN REDUCTIONS

Ing-Marie Gren

On the basis of the evidence presented thus far, it seems reasonable to conclude that, for a 50 per cent reduction strategy in nutrient loads, benefits cover the costs. But what would the reduction be if the countries tried instead to maximize net benefits – benefits minus costs? Would there be a difference between a coordinated approach, where total net benefits are maximized and a policy where each country considers only its own net benefits? In principle, we can say that the answer to this question is yes. There are differences between the coordinated and uncoordinated policies, if there are links between the countries. These links imply that domestic action taken that reduces/increases nutrient loads affects the quality of waters in other countries. The purpose of this chapter is to, by means of simple game theoretic tools, analyse and calculate nitrogen reductions and net benefits under different cooperative and non-cooperative scenarios.

There is a large body of literature, mainly theoretical, with a focus on the efficient provision of an international environmental public good (see Barett, 1992; Germain et al, 1995; Hoel, 1992; Mäler, 1991 and 1993; Kaitala et al, 1992 and 1995; Markovska and Zylicz, 1996; Chander and Tulkens, 1992). Only very few of these studies are empirical applications to concrete international environmental concerns (Kaitala et al, 1992 and 1995; Mäler, 1991 and 1993; Markovska and Zylicz, 1996). Common to this literature is the inclusion of only one type of pollutant-transport medium, which usually also constitutes the only spillover effect between countries. The main focuses of these studies can be roughly divided in to two classes related to the provision of an international environmental good: the determination of whether or not cooperation takes place (Mäler 1991 and 1993) and the analysis and calculation of appropriate cost-sharing schemes (Kaitala et al, 1995; and Markovska and Zylicz, 1996). Exept for the Markovska and Zylicz (1996) study, which is applied to nitrogen loads to the Baltic Sea, all of the studies are concerned with acidification from sulphur emission in Europe and neighbouring countries. All of the sulphur studies also consider the dynamics of the environmental damages.

As with the Markovska and Zylicz study, a simplifying assumption is made that environmental benefits and emission reductions occur in the same period. This is explained by the current insufficient knowledge of the nitrogen dynamics in the Baltic Sea and an associated lack of data. A similarity between all the studies is the inclusion of marine water transports of nitrogen as one type of environmental benefit link between countries. The difference in the analysis in this chapter is the recognition of the fact that most pollutants are transported, not only by one environmental media, but by several, such as air and water. In a region such as the Baltic Sea drainage basin, this is likely to imply at least two classes of environmental links between the countries: air transfer and deposition and marine water transports. An attempt is made in this study to analyse and quantify the different outcomes from a non-cooperative game for nitrogen reductions, including both types of environmental spillovers. The non-cooperative framework is applied because of the current impossibility of making binding agreements among the Baltic Sea countries. A unique feature is the existence of relatively comprehensive data sets on nitrogen-media transports and the costs and benefits of nitrogen reductions at the scale of Baltic Sea and its drainage basin (Gren et al, 1997). In other respects, the purpose and methodological approach in this study is similar to that of Mäler (1991) and (1993).

The chapter is organized as follows. First, the model analysing the outcomes under different solutions is presented. Next, nitrogen transports and costs of nitrogen-reduction measures are briefly described. The calculations of nitrogen reductions and net benefits are then presented. Unless otherwise stated, all results and data referred to can be found in detail in Gren (1998).

THE ANALYTICAL THINKING

Since there (currently) is no Baltic Sea public body which can enforce pollutant reductions among the Baltic Sea countries, a substantial improvement in the biological condition of the Sea relies on the individual country's ability and willingness to make voluntary agreements. One way of approaching the potential for agreement is to look at the national gains and losses from cooperation and non-cooperation. This can be more or less simple, depending on the country net gains in the two situations. In order to illustrate this, let us imagine only two countries: Sweden and Poland, which contemplate cooperation (*C*) or non-cooperation (*NC*). Assume further that they make their decision simultaneously. Each outcome – only one country cooperates or both cooperate or not – generates certain net benefits for each country. This simple game is presented in a strategic form in Table 8.1. The numbers in the table are hypothetical.

The hypothetical numbers in each cell reflect the net benefits to Sweden and Poland respectively. For example, in the *C, NC* cell the net benefits to Sweden are six and to Poland zero. We first start by investigating the likely choice of Poland. What would Poland choose if Sweden decides to cooperate? In Table 8.1 we see that Poland can then gain ten if it cooperates and

Poland

		C	NC
Sweden	C	10, 10	6, 0
	NC	0, 6	1, 1

C = cooperation
NC = non-cooperation

Table 8.1 *Game Illustration A*

zero under non-cooperation. As the net gains are larger under cooperation, the choice will be to cooperate. In a similar way, we can see that Poland would also choose to cooperate also if Sweden decides not to cooperate. Obviously then, the Polish choice is unaffected by the Swedish choice.

Let us now turn to Sweden and ask the same question. What will Sweden do if Poland decides to cooperate? Obviously, the gains for Sweden are higher under cooperation. We can also see that gains from cooperation are higher even in the case of Poland deciding not to cooperate. Thus, since both countries decide to cooperate regardless of the other country's choice, cooperation is the dominating strategy and will be the outcome of this simple game. The policy conclusion from this simple game is then that if the net benefits for the Baltic Sea countries under different outcomes are as presented in Table 8.1 – that the choice between cooperation or non-cooperation is independent of what other countries do – the issue is 'only' to inform all partners of the distribution of net benefits. The cooperative outcome, which generates the maximum net benefits for all countries will then occur. It will be in the self interest of each country to cooperate. Since there is not full information on net benefits under different pollution-reduction strategies for all of the Baltic countries, we cannot rule out the possibility of the neat situation illustrated in Table 8.1. However, it may very well also be the case that non-cooperation, with less net benefits for each country will be the outcome (Table 8.2).

Following the same procedure as before, we note that *NC*, *NC* will be the outcome. When Poland decides to cooperate, the best choice for Sweden is *NC*. The same holds for Poland when Sweden decides to cooperate. When one of the countries makes the choice *NC*, the largest gains for the other country will also be not to cooperate. However, recall that we assumed a so-called one-shot game – there are no communications before the game is played. We would expect that a repeated game or a game allowing for pre-communication would generate a cooperative outcome since both countries gain from cooperation as compared to non-cooperation. The policy issue in this situation could then be the need for the institutional provision of information and negotiation opportunities.

Poland

		C	NC
	C	5, 5	0, 6
Sweden			
	NC	6, 0	1, 1

Table 8.2 *Game Illustration B*

The environmental problems underlying the two games A and B can be classified by so-called mutual environmental relations. That is, the strategies of both countries affect each other. Another class of transboundary environmental problems is characterized by more or less unilateral environmental linkages. A typical such case is international rivers. The pollution from upstream countries affects the environmental quality of downstream countries, but not vice versa. In this stylized case, when the upstream country gains no domestic environmental benefits from pollutant reduction, cooperation implying pollutant reductions in both countries will never occur, unless the downstream country compensates the upstream country for its losses. We may think of the Baltic Sea nitrogen pollution in a similar way, since the marine-water currents transport nitrogen upwards along the west side of the Baltic Sea, but not in the other direction, because the northern Baltic Sea water basins function as nitrogen sinks. Poland could thus be regarded as an 'upstream' polluting country and Sweden as a 'downstream' polluted country. Such a situation is illustrated in Table 8.3.

In the simple illustration in Table 8.3, it is assumed that there are no impacts on the Polish coasts from Swedish nitrogen reductions. Cooperation over nitrogen reductions by a certain amount further implies a net loss to

Poland

		C	NC
	C	10, -3	2, 0
Sweden			
	NC	7, -3	-3, 0

Table 8.1 *Game Illustration C*

Poland, because of the very small impacts on its domestic coast. When there are no nitrogen reductions – the *NC, NC* cell – Sweden faces a net loss of -3, while the are no net effects for Poland. However, when calculating the sum of net benefits for *both* countries we note that the maximum net benefits occur where both countries cooperate, which amounts to 7 (10-3). Thus, if both countries agreed on this solution, they would share the maximum net benefits, after the Swedish compensation of the Polish loss of -3. If such side payments cannot be made, a more likely outcome is *C, NC*, where Sweden makes nitrogen reductions and Poland does not. We than see that net benefits of 5 (7-2) are lost to both countries. The achievement of an optimal outcome in this type of game therefore requires side payments in addition to the information and institutional provisions. We cannot say which of the game illustrations A, B or C best describes current relations between the countries with respect to the costs and benefits from nitrogen reductions, which is an empirical matter. In the next sections we will attempt to identify links and to calculate the benefits and costs when there are not only two choices as in the preceeding examples, but a continuum of nitrogen-reduction options for all nine countries in the Baltic drainage basin.

NITROGEN TRANSPORTS AND THE COSTS AND BENEFITS OF NITROGEN REDUCTIONS

This section contains a very brief presentation of the data sets required for the calculation of net benefits under different solutions. We rely very much on the nutrient transport data presented in Chapters 2 and 5 and also apply the benefit estimates from Chapter 6. In addition, information is required on the transport of nutrients between countries in the different Baltic Sea basins. Admittedly, although there are models of marine nutrient transports between Baltic Sea basins (see Chapter 5), there are no data on transports between countries in each basin. This is of particular importance for the Baltic Proper, which is shared between all countries other than Finland. Therefore, very arbitrary assumptions on transports between countries have been made in this chapter in order to illustrate the impacts of different coordinated and uncoordinated policies.

Estimates of the value of lower eutrophication in the Baltic Sea have been presented in Chapter 6. The results indicate a total willingness to pay of about SEK 31,000 million per year. As is common with many methods for estimating the monetary value of changes in the supply of an environmental good, only one change in the supply is considered. This makes it very difficult to trace marginal-change benefits between the initial supply and the provision assumed in the valuation scenario. Such marginal benefits are just what are required for the purpose of this analysis. A linear relation between nitrogen reductions and benefits has therefore been assumed, which implies a constant environmental marginal benefit, and further, that this is the same for all regions. The marginal benefit is obtained simply by dividing total benefits by 500,000 tons of nitrogen, which is the 50 per cent decrease in total nitrogen load to the Baltic Sea. The estimated marginal benefit is then SEK 62/kg nitrogen reduction.

In Wulff et al (1990), nitrogen marine transports between three major Baltic Sea basins – the Baltic Proper, Bothnian Sea, and Bothnian Bay – have been calculated. According to the results, 20 per cent of the nitrogen load to the Baltic Proper is transported to the northern basins – the Bothnian Sea and Bothnian Bay. These northern basins are shared by northern Sweden and Finland. Due to the marine streams, there is no transport from these basins to the Baltic Proper. Therefore, it is assumed that Finland and northern Sweden receive 20 per cent of all the nitrogen load from all other regions. Further, the loads from northern Sweden and Finland imply marine environmental impacts on these regions only.

Unfortunately, the basin-nutrient transport calculation does not contain any estimates of the transports between regions within each basin, which means that there are no estimates of marine nitrogen transports between the eight countries sharing the Baltic Proper basin. Therefore, arbitary assumptions have been made that are based on certain information on the coasts of the countries. When the coastlines contain islands and vegetation, more nitrogen impact occurs on the coast of the emitter country and vice versa. It is assumed that the coasts of Poland and Lithuania are 'open', implying that most of the nitrogen transported from these countries is dispersed in to the Baltic Proper. It is therefore simply assumed that all the nitrogen that is received by the Baltic Sea from these countries is equally divided between all the Baltic Proper countries. For the remaining countries it is assumed that one third of the impact occurs on its own coast, while the remaining part is equally divided across the other countries. In order to calculate the impacts on the Baltic Sea from different nitrogen reduction measures, information is needed on the share of the nitrogen emissions that reach the Baltic Sea. With the exception of sources with direct impact on the Baltic Sea, this share is determined by several factors: share of air emissions that reach the Baltic Sea and other regions, leaching from land and retention during the water transport from the emission source to the coast. The marginal costs of

Table 8.4 *Calculated Domestic and Exported Marginal Benefits, SEK/kg Nitrogen Reduction*

Region	Marginal benefits	
	Domestic	Export
Denmark	31	31
Finland	37	25
Poland	7	55
Germany	19	43
Russian Federation	12	50
Estonia	19	43
Latvia	19	43
Lithuania	19	43
South Sweden	19	43
North Sweden	37	25

reductions in the load of nitrogen to the coast for different measures de-pend on these factors and the cost of the nitrogen reduction and mitigation measures at the source (Table 8.4).

The estimated domestic marginal benefits vary between SEK 6/kg ni-trogen reduction and 37/kg nitrogen reduction. The numbers tell us that a country will not implement nitrogen-reduction measures when its mar-ginal costs are higher than the marginal benefits. Thus, Poland will under-take nitrogen-reduction measures with marginal costs less than SEK 7/kg nitrogen reduction. Recall, however, the arbitrary assumptions on marine-nutrient transports. Since Poland accounts for about 33 per cent of the total nitrogen load and further receives relatively low benefits from nitrogen reductions, relations between the countries is most like the game illustra-tion C illustrated earlier. That is, Poland can be regarded as an 'upwind' country, where the largest gains from nitrogen reductions are obtained by other Baltic Sea countries. It is thus expected that Polish nitrogen loads and net benefits will be significantly affected by the type of solution that is implemented.

ESTIMATED NET BENEFITS UNDER ALTERNATIVE BARGAINING SOLUTION CONCEPTS

In this section, results from the estimation of optimal net benefits under different bargaining solutions as analysed earlier are presented, for exam-ple 'full cooperation', 'Pareto-efficient' outcome and 'Nash equilibrium'. The full cooperation outcome is underpinned by the maximized net benefits rule that ensures the total costs of nitrogen abatement are minimized. A Pareto-efficient outcome would be one in which no individual participat-ing country could be made worse off under the bargaining arrangements. The Nash equilibrium implies a lack of cooperation and each country con-centrates solely on minimizing its own costs. In addition, net benefits of alternative interpretations of the ministerial agreement on 50 per cent ni-trogen reduction to the Baltic Sea are presented. The algorithm applied in all calculations is found in Brooke et al, 1992.

Optimal nitrogen reductions

From the analytical presentation and the description of nitrogen loads and transports, together with marginal costs of nitrogen-reduction measures, we would expect Poland to be a net loser in a full cooperative solution. The reasons are its large share of nitrogen loads, large exported share of marine environmental benefits, and low-marginal costs of its own nitro-gen reductions. This expectation is confirmed by the presentation of re-sults in Table 8.5.

Table 8.5 indicates that overall there is only a small 0.1 per cent differ-ence between total net benefits in the full cooperative solution and Pareto solution. However, due to the increase in nitrogen loads from Poland, the net benefits are reduced, between 2–4 per cent, for all other countries. The Nash solutions imply a larger decrease in total net benefits, SEK 10,937 million

Table 8.5 *Nitrogen Reductions, Percentage, and Net Benefits, (Millions of SEK) for Different Regions Under Alternative Solution Concepts*

Region	Max. net benefits % N red.	Net ben.	Pareto efficiency % N red.	Net ben.	Nash equilibrium % N red.	Net ben.
Denmark	40	1615	40	1575	28	183
Finland	31	1984	31	1945	14	28
Poland	53	–343	51	0	14	75
Germany	17	1324	17	1285	8	271
North Sweden	37	2221	37	2179	52	437
South Sweden	36	1056	36	1015	17	213
Russia	31	1108	31	1068	13	31
Estonia	34	1313	34	1274	33	37
Latvia	52	1240	52	1200	39	125
Lithuania	53	1108	53	1068	46	75
Total	**41**	**12,628**	**40**	**12,611**	**22**	**1691**

or 85 per cent, in comparison to the full cooperative solution. All countries, with the exception of Poland, gain from cooperation.

The total nitrogen reductions under the full cooperation and the Pareto-efficient solutions are almost the same, approximately 40 per cent. Under the Nash equilibrium, the total nitrogen decrease is lower and corresponds to 20 per cent of the initial load. Only Finland and Estonia have a higher nitrogen reduction in the Nash equilibrium than under the other solutions. The larger reduction in Finland is explained by the fact that, under cooperation, the country creates net benefits from its own nitrogen reductions in other countries, due to the influence of marine-water currents. This is also true for the northern Swedish region, but unlike Finland, this region has a relatively large domestic benefit gain from measures reducing air emissions. The Estonia reduction in the Nash equilibrium is explained by the unique large areas of natural wetlands which, under a cooperation solution, act as nitrogen sinks for air deposition from all other countries. The benefits from using these wetlands as filters is greatly reduced under the Nash non-cooperative equilibrium, since the air emissions within the Estonia border are relatively small.

Net benefits from a 50 per cent nitrogen reduction

According to the ministerial declaration, the load of nutrients to the Baltic Sea shall be reduced by 50 per cent (HELCOM, 1993). It is, however, by no means clear how this agreement is to be interpreted. In principle, there are two options: the total load is reduced by 50 per cent, or each country reduces its own load by 50 per cent. The estimated maximum net benefits from the 50 per cent nitrogen reduction under these two alternative interpretations are presented in Table 8.6.

Table 8.6 *Nitrogen Reductions, Percentage, and Net Benefits (Millions of SEK) from a Nitrogen Reduction by 50 Per Cent*

Region	Total reduction by 50 %		Country reduction by 50 %	
	% N red.	Net ben.	% N red.	Net ben.
Denmark	27	1495	50	-217
Finland	48	1671	50	-631
Poland	63	-3787	50	-1967
Germany	17	851	50	-16,600
North Sweden	41	2487	50	-316
Russia	47	1202	50	-614
Estonia	47	1092	50	129
Latvia	64	1086	50	166
Lithuania	63	547	50	123
South Sweden	40	579	50	-1811
Total	**50**	**7180**	**50**	**-21,738**

The first interpretation gives a total net benefit of SEK 7180 million, which is 43 per cent lower than under the full cooperative solution. The only region with negative net benefits is Poland, whose losses are ten times as high as compared with the full cooperative solution outcome. Although most countries gain from cooperation, north Sweden and Russia would particularly favour a target of 50 per cent reduction in total nitrogen. However, the total net benefits when each country reduces its load by 50 per cent is approximately SEK -21,738 million. This large total net loss is due to the very high increase in costs for German reductions, which, in turn, are due to the high marginal of costs of reductions in air emissions and its large share in air deposition. The results in Table 8.6 also indicate that a majority of the countries face negative net benefits from uniform country restriction in the load of 50 per cent.

SUMMARY AND DISCUSSION

The main purpose of this study has been to calculate the net benefits from reductions in the load of nitrogen to the Baltic Sea under alternative reduction strategies. The complications introduced by the inclusion of several environmental nitrogen transport media in the analysis constitute an important challenge when estimating net benefits. The load of nitrogen consists of direct discharges into the Baltic Sea, loads which are deposited on land and transported by water flows to the Baltic Sea. The direct loads include air deposition and the discharges of sewage treatment plants and industry located at the coast. The water-borne transports originate from the upstream deposition of nutrients related to agricultural, residential and industrial activities, and air borne emissions.

Three different bargaining solution concepts were analysed: maximum net benefits, Pareto efficiency, and Nash equilibrium. Under the maximum net total benefits rule, total costs are minimized – costs of nitrogen reduction and environmental costs of nitrogen loads. The Pareto-efficient solution is defined as a situation in which no region is made worse off than they were prior to any reduction; and the Nash non-cooperative equilibrium implies that each country treats other regions' actions as given and minimizes its own total costs. The difference in the outcomes between these three solutions is determined by the two nitrogen linkages between the Baltic regions; the marine transports of nitrogen and the air transports of nitrogen.

The empirical calculations indicate a negligible difference between the maximum total net benefits and the Pareto-efficient solutions. The maximum total net benefits solution can, however, only be secured through the sanctioning of side payments to Poland. The Nash equilibrium implies a reduction of 16 per cent in total net benefits. The reason for this is the reduction in Polish nitrogen loads that have relatively large environmental impacts on the other regions. Under the maximum net benefits and Pareto-efficient solutions, the total nitrogen load decreases by 40 per cent, and under the Nash equilibrium the load decreases by 33 per cent. If implemented, the current ministerial agreement for a 50 per cent nitrogen reduction would yield, in total, net benefits that were approximately 40 per cent less than the net gain that would be achieved with the maximum total net benefits rule. If, however, the 50 per cent nitrogen-reduction agreement is imposed on each region, there is a large total net loss. This is due to the very high increase in costs for German nitrogen reductions.

It is important to note that these estimates are sensitive to several assumptions. For example, sensitivity analysis in Chapter 4 shows that total costs of nitrogen may be doubled when reduced-capacity of low-cost measures are assumed. Another crucial assumption concerns the linkage of nitrogen reductions to benefits as measured in monetary terms. Recall also the arbitrary measurement of marine nitrogen transports between countries. A change in these transport parameters might result in considerable changes in the allocation of nitrogen reductions and net benefits. However, although the numerical results must be interpreted with great caution, the results of this study indicate that: 1) the linkages between regions provided by marine nitrogen transports are important, while the linkages via air transport seem to be of less significance; 2) all regions would gain from giving side payments to Poland, as compared to a Nash solution; and 3) the current ministerial agreement for a 50 per cent nitrogen reduction implies lower total net benefits as compared with all three solutions.

The bargaining simulation results are also potentially likely to be influenced by the factors excluded from this analysis. One such factor is the neglect of the wider environmental benefits associated with nitrogen reductions. For example, the construction of wetlands implies further environmental benefits via the provision of increased biodiversity and also recreational/amenity gains. The inclusions of such 'extra' environmental benefits would increase nitrogen reductions for all regions, where these benefits are positive. Another important issue is related to the simple characterization of nitrogen reduction utilized in this study. A more precise measure would have included not only the direct costs of the reduction activities, but also their general equilibrium impacts. As shown in Chapter 9, structural impacts of nitrogen-reduction policies can be large, but the net impact on GDP is small.

This study, like many other similar studies, applies the unrealistic assumption of perfect compliance. That is, it is assumed that nitrogen-reduction agreements are implemented in each country at no cost. In Gren and Brännlund (1995) it is shown that the cost of enforcing a nitrogen-fertilizer charge may correspond to 25 per cent of total costs. Since there are large institutional differences affecting environmental policies between the countries, the enforcement costs are also likely to differ significantly (Eckerberg et al, 1996). The inclusion of enforcement costs would thus affect, not only total net benefits, but also the allocation of nitrogen reductions between different regions. However, in spite of these drawbacks, the following question is of some interest. If the results indicate gains from pollutant reductions under the Nash equilibrium, why are these strategies not implemented by the countries? Given sufficient correspondence between the actual decision situation and the one modelled, we would expect the small net gains under Nash-equilibrium conditions to be realized in practice. Obviously, there is a difference. There are at least three explanations for the relatively high net gains under Nash-equilibrium conditions in this study: incorrect nitrogen transport matrices, high benefits and low costs of nitrogen reductions estimations, and ill-informed decision makers; each of these is briefly discussed below.

Only one source for the air borne nitrogen transport and marine water transport data has been used in this study. The quantification of the marine transport function is especially sensitive because it conditions the allocation of inter-country net gains and losses. The monetary benefit estimates applied in this study are also far from precise. A reduction in the benefits per unit nitrogen reduction by one-half, for example, implies a decrease in net benefits by about 66 per cent. Although decision makers may be well informed about the costs associated with implementing nitrogen-reduction strategies, a comparison of all options and their costs is required for the identification of cost-effective solutions. The first attempt to make a systematic investigation and comparison of various nitrogen-reduction options in the Baltic Sea countries was in Gren et al (1997). It is therefore quite likely that the information necessary for making cost-effective nitrogen reductions has not been available to decision makers and others.

In spite of our focus on countries within the drainage basin only, a maximum total net benefits solution might not be achieved without compensation to the net losers. Currently, the polluter-pays principle adopted by the OECD countries is an impediment to the use of side payments. One important concern is related to the disincentives created for pollution abatement in countries that become known as candidates for receiving payments. If side payments cannot be made, an alternative 'best' outcome may be the minimization of total costs without making any region worse off.

9 THE EFFECTS OF IMPLEMENTING MARKETS FOR EMISSION PERMITS NATIONALLY VS REGIONALLY

Åsa Johannesson and Paula Randås

INTRODUCTION

Thus far, there has been no consideration of the structural impacts or indirect costs on the national economies imposed by policies aimed at reducing nutrient emissions. In the preceding chapters, only the direct costs borne by the affected sectors were included. The purpose of this chapter is, there-fore, to compare the total effects, including direct and indirect costs, of a nitrogen permit trading market on four regions – the Baltic States, Poland, Denmark and Sweden. Nitrogen emission reduction is considered under two different policy approaches. The two cases analysed are reductions under-taken individually by the countries, and the cooperative solution when the countries combine to minimize the costs of the reductions for the region as a whole. In both cases the policy instrument used is a market for tradable emission permits, either one in each country or one for the whole region.

The main difference from the simulations carried out in Chapter 8 is the consideration, not only of the direct costs faced by firms' nitrogen-reducing activites, but also of the indirect effects on other sectors. By implementing an emission-reduction policy scheme, the relative prices of commodities will change. Products that have nitrogen as an input will become more expensive relative to other products. This will affect the whole economy, and thereby the production and consumption decisions. In order to capture all these effects, a computable general equilibrium (CGE) model is the most appropriate method to use.

We focus on nitrogen as the effluent because, as discussed in Chapter 5, it is currently the limiting nutrient in the largest part of the Baltic Sea. In our model the reduction of nitrogen is achieved through a permit market, and not through taxes or standards. Here the permit market is regarded simply as an instrument to reduce emissions. The explicit choice between different policy instruments has been analysed in several papers (see for example Howe, 1994, for an empirical analysis and a theoretical review). The initial distri-bution of permits is important for the outcome on the common market, since alternative allocation schemes for the permits will affect the incomes or costs for the countries or firms involved. There are several articles that consider the initial distribution of permits and the rules for such a arrangement (for example, Bertram 1992; Tietenberg, 1995).

In a multi-country setting with transboundary pollution, one important aspect of the initial distribution is the possibility of making transfers to countries that would otherwise be inhibited by high costs and thus unwilling to cooperate. Barett (1992) shows that as the number of countries taking part in an international agreement increases, the harder it will get to reach an agreement without transfers. This has to do with the fact that the more countries that are involved, the greater the differences in net gains between them will be. International environmental policy can also be modelled as non-cooperative games as shown in Chapter 7. Coordinated reduction policies are usually compared with the Nash equilibrium, where each country is concerned only with its own costs and benefits and disregards impacts on other countries from its own actions.

In the case of the Baltic Sea, the amount of nitrogen emissions to be reduced is specified in the Baltic Sea Declarations, and the question is rather how to obtain these reductions and what the costs of these measures will be. The benefits from the decreased pollution will be the same regardless of which policy instrument is chosen, and therefore for our purpose, the important part is the cost of reductions, not the benefits. Thus, in our model no damage functions are specified. Instead, we are concentrating on what the effects on the economies from reducing their emissions will be. Our base case, business as usual in 1991, can possibly be interpreted as a non-cooperative Nash equilibrium, to be compared with different forms of cooperation.

There are several CGE-models focusing on different aspects of environmental policy. Effects of unilateral reductions in greenhouse gas on a single country are analysed in Bergman (1991), Stephan et al (1992), Whalley and Whiggle (1991). In Burniaux et al (1992) a global, multi-regional model is developed to analyse the effects of curbing CO_2 emissions. As an analytical tool, our model is close to Bergman (1995) and also related to Robinson et al (1994). The main difference is that we do not focus on one, but several countries polluting a common resource, and on the effects on these countries from coordinating their emission reductions. The model used for this analysis is a CGE-model for six countries, each with seven producing sectors and one representative consumer. The emission reduction is implemented through a permit market, which has been simulated both as separate national markets and as a common regional market. Unless otherwise stated, all data and results referred to are found in Johannesson and Randås (1995).

MODEL AND DATA

In principle, the approach we use is a static multi-country CGE-model, originating from a single-country model by Bergman (1995). The countries included are identically modelled, and differ only with respect to the data representing each of them. Much of the structure follows a standard CGE-approach, where agents are price takers, which implies that no single firm can influence market prices by changing its supply of products. The representative consumer is further assumed to maximize utility from market goods and firms to maximize profits, subject to budget and technology constraints. Perfect competition is assumed for all markets, in the sense that prices are fully flexible and thereby market clearing is assured for all goods and factors of production. The factor endowment is fixed for each country, as are the aggregate investment expenditures. The countries are small in that they can not affect the world market price of goods. An imported good is modelled in terms of fixed Leontief-input co-efficients, which means that the inputs are used in fixed proportions for producing outputs so that there are no substitution options between imports and other inputs. The goods produced for the domestic market can, however, be substituted for export, through a constant elasticity of transformation function with the elasticity set to -0.5. The balance of trade is exogenously given. One drawback with the specification of the model is that the price of labour is assumed to be fully flexible, and hence unemployment is not part of the model. Incorporating unemployment could affect the results, especially for the eastern states. However, for simplicity, this possibility is not used here.

In Bergman (1995), a framework was developed for considering the possibility for firms to abate their emissions as an alternative to buying emission permits. We have expanded this possibility and use the marginal costs of abating nitrogen emissions in the Baltic Sea region that have been presented in Chapter 4.

There are seven producing sectors in the model, out of which four are nitrogen emitters. These sectors are agriculture, the nitrogen industry, sewage and transport. The 'clean' sectors are fertilizers, non-nitrogen industry and services. Each sector produces a specific good and uses inputs from imports, domestically produced intermediates, labour and capital, and in the case of the emitting sectors, a certain amount of nitrogen emissions. In addition to capital and labour, the agricultural sector also uses fertilizers as a substitutable input.

A large amount of data is required in order to simulate the results presented in this chapter. These have been collected from a number of sources and arranged to fit into the structure of the model. It should be noted that there have been difficulties in obtaining all the relevant data, in particular for the east European countries. As a consequence of these difficulties, figures for different years have been used, but as close to 1991 as possible. This section presents the data and its major sources. The model is constrained to six countries surrounding the Baltic Sea in order to limit the amount of data gathering and processing, and as a matter of simplification. The countries included in our analysis are Sweden, Denmark, Poland, Estonia, Latvia and Lithuania. One objective of this choice was to cover a large share of the emission of nitrogen to the Baltic Sea, and with this selection, two-thirds of the emissions are included in the model. For simplicity, we use an aggregate of Estonia, Latvia and Lithuania in the model. This aggregate will from now on be referred to as the Baltic States. Another objective was to use both east and west European countries, as it is likely that reducing the nitrogen emissions will affect these countries in different ways.

The production in all countries is divided into four polluting and three clean sectors. The four that emit nitrogen are agriculture, the nitrogen industry, transport and sewage. The three remaining sectors are non-nitrogen industry, service and fertilizers. The relative size of these sectors in the different regions is presented in Table 9.1.

In order to carry out CGE-modelling a description of the goods transported between different sectors and for consumption and exports, a so called social accounting matrix (SAM) is needed for all countries. Government services are excluded from the analysis in all countries, because the Swedish SAM that we use only includes producing sectors. The nitrogen emissions from this sector are negligible in Sweden and not available for the other countries. As can be seen from Table 9.1, there are large differences between the countries both in the relative size of the sectors and in the absolute level of production. For example, the GDP of the Baltic States is just a bit more than 1 per cent of the Swedish GDP.

SAMs have been available for two of the countries, namely Sweden and Poland, and were provided by the Swedish National Institute of Economic Research and the Statistical Office in Poland respectively. We then constructed the SAMs for Denmark and the Baltic republics by adjusting the sector specification and the size of the GDP of the Swedish and Polish SAM to fit the economic structure of Denmark and the Baltic States. However, as a result of this, the same production technology, in terms of the use of intermediates and factors, are assumed for Denmark and the Baltic States as for Sweden and Poland respectively.

The sewage sector is included in the model as a way to incorporate the emissions from the household sector. A large part of the production of sewage is not directly demanded as an intermediate or direct consumption, but is merely a fixed amount of sewage activity provided to the economy. The result is that the production of sewage is relatively fixed and is only marginally affected when the production cost and output price increase. Thus, the sewage sector can be interpreted as the amount of abatement related to the level of household emissions that takes place in the base case. This abatement

Table 9.1 *Sectors Shares in Percentage of Total Output (Exclusive of Governmental Services), and GDP (Millions of SEK), 1991*

Sector	Sweden	Denmark	Poland	Baltic States
Nitrogen emitting sectors:				
Agriculture	2.8	5.3	11.7	21.0
Nitrogen industry	27.3	15.1	38.8	24.2
Sewage	0.4	0.3	0.3	0.2
Transport	5.6	10.3	5.2	9.2
Non-emitting sectors:				
Fertilizer	0.1	0.1	0.5	2.4
Non-nitrogen industry	32.0	18.7	30.7	12.1
Service	31.8	50.2	12.8	30.9
Output	100.0	100.0	100.0	100.0
GDP in millions of SEK	**1,192,038**	**580,441**	**305,091**	**16,755**

Source: Cited in Johanesson and Randås (1995)

capacity can then be expanded via innovations in new abatement technologies. Nitrogen is one of the major ingredients in fertilizers. As a simplification, we specified fertilizer as a substitutable factor in agricultural production in order to be able to tie the nitrogen emissions directly to this input. This possibility is not used here though, as it turned out that the results generated with this specification of the model mainly came from very large substitutions to capital from fertilizers. We interpreted these results as rather unrealistic, as the only part of the capital that can be substituted against fertilizer is land. However, land is not specified in our model as a factor on its own, due to lack of data. As a result, the emissions in agriculture are also considered to be output related. For our model, the water and air emissions of nitrogen have been aggregated and related to the four emitting sectors. Table 9.2 provides an overview of the quantities of emitted nitrogen by different sectors in the four regions. It should be noted that we assume in this application, that the Baltic Sea is a homogenous body of water, and the effects of the emissions of nitrogen is the same throughout the whole region, regardless of where the emissions take place.

As Table 9.2 suggests, the emission levels differ both between countries and sectors. In general, the agriculture and sewage sectors are the major polluters. The major source of nitrogen within agriculture is the use of fertilizers and the main part of the sewage emissions originate from households. The NO_x emissions originating from the transport sector are consequences of the use of fuels. The nitrogen emissions in the industrial sector are byproducts of different production processes. Some of the largest nitrogen emitters within the whole industrial sector are the pulp and paper industry, the food industry, and for the eastern countries, the chemical industry. Since the emissions of nitrogen are residuals of production, they can be regarded as inputs into the production process. The emission coefficients are calculated as the initial emissions for each sector in relation to initial total sector output. From Tables 9.1 and 9.2 it follows that these coefficients vary greatly between the different countries as well as between sectors, which can be confirmed by studying Table 9.3.

The eastern states have emission co-efficients that are generally several times larger than in the west, since Poland and the Baltic States emit a lot more relative to their production. The co-efficients for the Baltic republics are higher again than those for Poland.

Table 9.2 *Emission of Nitrogen by Sector and Country (Millions of Tons, and Per Cent)*

Sector	Country Sweden		Denmark		Poland		Baltic Rep.	
	tons k	%	tons k	%	tons k	%	tons k	%
Agriculture	32.4	43	41.5	69.2	100.9	51.2	27.2	52.3
Nitrogen industry	8.6	11.5	4.1	6.8	28.5	14.5	4.6	8.8
Sewage	29.3	38.9	11.0	18.4	64.5	32.8	19.1	36.7
Transport	5.0	6.6	3.4	5.6	3.0	1.5	1.1	2.2
Total	*75.3*	*100.0*	*59.9*	*100.0*	*196.9*	*100.0*	*52.0*	*100.0*

Source: Gren et al (1995)

Table 9.3 *Emission Co-efficients, Kg Nitrogen/SEK output*

Sector	Country			
	Sweden	Denmark	Poland	Baltic States
Agriculture	0.42	0.62	1.43	4.15
Nitrogen industry	0.01	0.02	0.12	0.61
Sewage	2.83	3.33	39.89	262.75
Transport	0.03	0.03	0.09	0.39

SIMULATION RESULTS

The initial situation in 1991 is used as the base or reference case. The emission reductions are implemented through a permit market, which is national or regional. When a common, regional market is enforced, the initial distribution of permits will have a large impact on the outcome. The allocation device used first is the principle of grandfathering, ie the permits are allocated according to the initial distribution of emissions, adjusted for the targeted reduction. In this scheme, the countries have exactly the same endowment of permits whether a regional or a national market is implemented. Thus, the biggest polluters will obtain the most permits, which in this model means the eastern European countries. In the last section, the size of the redistribution necessary to make a common market preferable to all countries is discussed and analysed.

National markets

There are two effects from implementing a permit market that drive most of the results. The first is that the marginal cost of production increases in the polluting sectors, since they now have to buy permits for all their emissions. When the relative prices change, this induces changes in the structure of private demand, export, and thereby, production. In addition, through the use of intermediate inputs, the marginal cost of production tends to also increase in the clean sector. The current account for each country is fixed to the 1991 level, as exogenous forces determine the size of the current account surplus or deficit. When the domestic price increases, the firms choose to export less and sell more on the domestic market. With a price level too high to be competitive on the world market, the factor prices, ie the price of capital and labour, start to fall. In equilibrium, these prices have fallen enough to establish a general equilibrium in which the state of the current account is maintained.

The second effect of implementing a permit market is that an abatement activity begins to demand resources. Since the factor endowments are assumed to be fixed in the economies, the competition for these factors increases. The production possibilities for the rest of the economy decrease and at the same time, when profits are created from the abatement activity, consumer income, and thereby private demand, increases. The abatement activities therefore promote increased costs of labour and capital used in producing good and services.

When national permit markets are implemented, the results indicate that the reductions will, in many aspects, be most costly in the Baltic States. In terms of utility losses, the country that will suffer the least from the restrictions is Sweden. In all countries large structural adjustments will be needed to achieve the emission reductions required. For reduction levels of 20–30 per cent, the output in most production sectors decreases in all countries. Increasing domestic prices combined with reduced consumer income, result in large reductions in private consumption and thereby also of the utility from consumption. The Baltic republics experience the largest reductions, followed by Poland. As can be seen from Figure 9.1, the utility loss is more modest in Sweden and Denmark.

In equilibrium, the impact on the total costs of a marginal change in the restriction on the target of nitrogen reduction in each country, will determine the price of permits. According to the simulations, there is a large percentage difference between the countries with regard to the national permit prices for the same level of nitrogen reductions. This can be seen in Table 9.4. This reflects the differences in marginal cost between the countries that were also shown in Chapter 4, where only direct costs were included.

Within the western countries, the marginal cost of reducing nitrogen increases much more in Sweden than in Denmark for large overall reductions. This can be explained by the lower marginal cost of abatement in agriculture in Denmark, which gives this country a cheaper abatement possibility in one of the sectors emitting the most, 69 per cent, of total emissions. In Sweden, the second cheapest technology can be found in the industry sector, from which only 12 per cent of the total Swedish emissions originate. The marginal cost of abatement in the Baltic republics is constant, which reflects the fact that abatement takes place in the sewage sector throughout the programme.

When looking at the effects on consumption at low reduction levels, we first get the intuitive results that the demand for clean goods, the non-nitrogen industry and services, increase, while the demand is reduced for

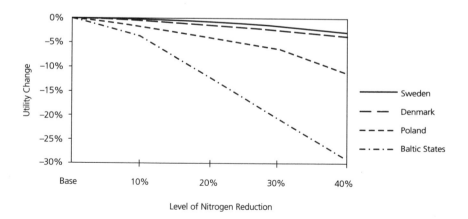

Figure 9.1 *Changes in Utility in Per Cents from Percentage Changes in Nitrogen Reductions*

Table 9.4 *Marginal Costs of Reducing Nitrogen Emissions in SEK/ton Nitrogen Reduction*

Reduction level	Country			
percentage	Sweden	Denmark	Poland	Baltic Republics
10	24,100	24,000	7000	7000
20	24,100	37,900	7000	7000
30	60,100	43,900	16,100	7000
40	139,300	43,900	29,900	7000

the polluting goods, with the exception of transport. This is a result of the higher production costs for the polluters, caused by the need to buy emission permits to be able to continue to produce. The increased domestic prices on goods from the polluting sectors reduce both private demand and the export of these goods, and hence production decreases. The reason that the demand for transport is not reduced is the small emission co-efficients in this sector. The domestic price of transport therefore increases only slightly in Sweden and Denmark, while in eastern Europe, it actually decreases due to reduced factor prices. For stricter emission levels, the demand for transport also falls in all countries but Poland. At the 40 per cent reduction target, total consumption is reduced most in the Baltic republics, 28 per cent, and least in Sweden, 2.5 per cent.

In sectors that do not emit nitrogen – the non-nitrogen industry and services – the domestic price falls. This is due to the fact that decreasing prices of labour and capital reduce the production cost in these sectors. These prices decrease across all reduction levels in all four countries. As a result, the private demand for the non-nitrogen products and services increases for small emission reductions. However, for larger emission reductions, demand starts decreasing in the Baltic States for both sectors and in Poland for the service sector. In Denmark, although demand falls, this is marginal compared to the base case. This fall in demand can be explained both by an income effect through reductions in the factor prices and decreased production possibilities. At high levels of nitrogen reduction, the abatement activities capture an increasing amount of the factors in the economies and hence less output can be produced. This in turn results in lower consumption possibilities in the clean sectors also and hence less production output. This in turn results in lower consumption possibilities in the clean sectors also. The reason why these effects are more pronounced in east Europe than in the west European countries is the much larger total amount of emissions in the east, which cause the abatement activity to take up a substantially larger share of the production factors there. Thus, the original production decreases more, which can be seen from Table 9.5. Total production is reduced by about 2 per cent in Sweden and Denmark, while in Poland and the Baltic republics the reduction is 6 per cent and 17 per cent respectively. Agriculture is one of the sectors most negatively affected in all countries. The reductions in production will also be significant in the nitrogen industry. The decreases in transport and sewage are generally much lower and amount to less than 2 per cent in all countries, the only

Table 9.5 *Percentage Change in Output from 40 Per Cent Country Nitrogen Reduction*

Sector	Country			
	Sweden	Denmark	Poland	Baltic Republics
Emitting sectors:				
Agriculture	−17.6	−11.1	−18.2	−23.3
Nitrogen industry	−4.9	−4.6	−8.6	−18.7
Sewage	0.5	1.7	0.1	1.6
Transport	−1.6	−1.6	0.7	−13.3
Non-emitting sectors:				
Fertilizers	−17.7	−12.1	−19.3	5.6
Non-nitrogen industry	0.6	−0.5	−0.4	−8.0
Service	0.0	−1.1	−3.1	−16.7
Total output	**−1.8**	**−2.1**	**−6.0**	**−16.6**

exception being the transport sector in the Baltic republics, where the production is reduced by 13 per cent.

Fertilizer production is somewhat different from the other sectors, since it is only used in the economy as a substitutable input in agriculture. The relation is not 1:1, though, because the fertilizer producers can also export their output at a fixed world market price. The production of fertilizers decreases for all levels of emission reductions in all countries, except in the Baltic republics, where it actually increases. The reason is that the price of fertilizers falls much faster in the Baltic States relative to other prices.

When stricter emission limits are implemented, the marginal production costs increase and the polluting sectors become less competitive relative to the world market prices. As a consequence, the factor prices begin to fall. This decrease is not as large in Sweden and Denmark, about 11 and 4 per cent respectively for both labour and capital, while in Poland and the Baltic republics it is around 30 and 40 per cent respectively. The main reason for the larger reduction in the eastern countries is that the polluting sectors represent a larger share of the total production and export of goods, and therefore a larger decrease in the factor prices is required to re-establish the state of the current account in these countries.

Regional permit market

The driving forces when a regional permit market is implemented are to a large extent the same as in the case of national permit markets. The marginal cost of production increases in the polluting sectors due to the need to buy permits, and changing relative prices induce changes in the structure of private demand, export and, thereby, production. Also, the abatement activity begins to require resources and due to the fixed factor endowments, the competition for these factors increases. The production possibilities for the rest of the economies then decrease.

However, another important driving force also exists in the case of a common permit market. Each country is given an endowment of permits equal to their initial amount of nitrogen emissions, and they can then export or import permits depending on the marginal cost of abatement in their economies. Thus, the export incomes and import costs from trading emission permits become an additional driving force in the simulations. When a country receives export incomes from selling emission permits, consumer income, and thereby demand, increase. The increased domestic demand results in increasing domestic prices through rising factor prices that result in decreased exports. Compared to the case of national permit markets, a smaller part of the current account will consist of goods from the producing sectors, as a new part will be generated by exporting permits. The opposite holds for a country importing emission permits.

Cooperation through the implementation of a regional market for nitrogen emission yields quite different results compared to the national markets case. For the region as a whole, the costs in terms of reduced utility are smaller under a common market. The equivalent variation (EV) – the income compensation required to keep the countries at the same utility level as before the implementation of the nitrogen emission targets – for all countries, amounts to approximately SEK 47,200 million under a common market, and about SEK 56,000 million for regional markets. The total costs, as measured in income compensation required to keep the countries at the same utility level as that prior to the change, are thus reduced by almost 20 per cent when regional markets are replaced by a common market.

However, the impacts on each country are not so clear-cut. At a 40 per cent nitrogen-emission reduction level, the consumers in Sweden, Denmark and the Baltic republics are all better off under the regional solution as the decrease in their utility is less then under national markets. However, Polish consumers should prefer separate markets as the opposite result is obtained here. This is illustrated in Figure 9.2.

These results are consistent for all reduction levels in Sweden and Denmark, but are not as stable for the other regions. Poland will prefer the common market for a 20 per cent reduction, and the national arrangement at

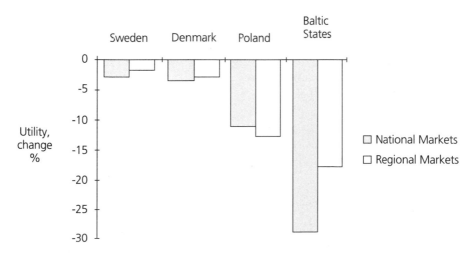

Figure 9.2 *Percentage Reduction in Utility from National and Regional Permit Markets at a 40 Per Cent Reduction Level*

all other levels. The Baltic republics should be willing to cooperate on a regional market in all cases except at the reduction level of 10 per cent.

In a perfectly competitive setting, it may seem strange that the introduction of trade in a good – permits for nitrogen emissions – actually reduces welfare in one of the trading countries. The explanation for this is that there are two effects at work. First, the consumer income is affected by the export or import of permits. Second, the whole economy is affected by the extra cost of reducing emissions. The permit price is now determined by the marginal cost of reducing emissions in the region as a whole, and it is not certain that the net impact of these two effects is positive in all the countries involved.

As mentioned earlier, emission permits are distributed according to the principle of grandfathering. This results in Poland and the Baltic republics obtaining rather large amounts of permits. Combined with their relatively low marginal cost for both abatement and reductions through changes in the production pattern, this makes these two countries net exporters of permits, while western states are net importers. This new export income is largest, as a share of consumer income, in the Baltic States, where it is between 4 and 9 per cent. The corresponding share is less in Poland, between 0.5 and 6 per cent. The import of permits constitutes a cost for Sweden and Denmark, but its share of consumer expenditure is rather small, less than or around 1 per cent. A comparison of the equilibrium permit prices is presented in Table 9.6, where it can be seen that the prices under a common market are somewhere between the higher and lower prices under national markets.

Thus, the regional market permit price makes the adjustment to the reduced emission levels easier for Sweden and Denmark, while the opposite is true for Poland and the Baltic republics. With a regional permit market, the largest emission reductions take place in the eastern states, as it is cheaper to reduce the emissions there. Compared to the case of national permit markets, a regional market will affect both the consumption and the production more in the eastern countries. Selling permits generates income to the consumers in Poland and the Baltic republics, with their initially very high emission levels. Given low abatement costs and low income level, the extra income generated from selling permits means quite a large extra injection of consumers' expenditure.

Table 9.6 *Marginal Costs of Nitrogen Emission Reductions in SEK/ton Nitrogen Under National and Regional Permit Markets*

Reduction level percentage	National permit market				Regional market
	Sweden	Denmark	Poland	Baltic republics	
10	23,500	23,400	6400	5100	6510
20	23,500	36,600	6200	4800	12,460
30	56,900	42,200	13,000	4500	22,890
40	123,500	42,200	20,700	4200	33,040

Consequently, consumption does not decrease as fast as in the case of national permit markets. The structural changes are larger, though. The high permit price changes the relative domestic prices quite a lot, and the tendency to choose clean goods instead of dirty ones is more apparent in the consumption pattern in the case of a regional permit market, than with national ones. Also, Polish consumers receive extra income from permit exports, but in their case this income as a share of total consumer income is smaller, and is not enough to compensate for the extra cost incurred from reducing their emission. At the 40 per cent reduction level, the decrease in consumption for Sweden, Denmark, Poland and the Baltic republics is about 2, 3, 11 and 13 per cent respectively.

As could be expected, prices increase less in Sweden and Denmark under a regional permit market than under national markets. Both these countries are net importers of permits, but the possibility of buying permits instead of employing expensive abatement technologies is still the dominating effect. Prices increase less and the relative prices change less, resulting in smaller overall changes in the consumption patterns as compared to the case of national permit markets. However, the reduction in total consumption is about the same in both countries. The regional market is clearly beneficial for producers in Sweden and Denmark, as the possibility of buying emission permits reduces boh the cost of production and the need to invest in abatement, as compared to the national permit markets. Both the reduction in the total production level and the structural changes are therefore smaller in Sweden and Denmark, which can be seen from Table 9.7. The fertilizer sector in the Baltic republics is one of those most affected if a regional market is used. Since the permit price and the factor prices are a lot higher on the regional market for the Baltic republics, this will increase the price of fertilizers. While in the national permit market case fertilizer production increases due to the increased export potential, it now decreases as much as 56 per cent at the 40 per cent nitrogen reduction level.

In all countries other than the Baltic republics, factor prices fall for all reduction levels. In the Baltic republics, factor prices increase substantially due to the large income from permit export. Consumer demand increases

Table 9.7 *Percentage Output Changes from an Overall 40 Per Cent Nitrogen Emission Reduction*

Sector	Country			
	Sweden	Denmark	Poland	Baltic Republics
Emitting sectors:				
Agriculture	-7.2	-9.8	-23.0	-31.9
Nitrogen industry	-1.9	-3.6	-13.1	-23.4
Sewage	-0.3	-0.3	-0.8	-1.6
Transport	0.1	1.6	-10.9	-27.6
Non-emitting sectors:				
Non-nitrogen industry	0.5	1.3	-5.0	-13.7
Service	-0.2	-0.2	-2.3	-0.6
Fertilizer	-7.2	-12.9	-32.5	-56.0
Total output	*-0.6*	*-0.7*	*-10.3*	*-18.1*

and exports decrease sharply. However, at the 40 per cent nitrogen-reduction level, the factor prices are lower than in the 30 per cent case. What
happens in the Baltic republics at this point is that the cheap abatement
technologies have been exhausted, and a relatively large part of the country's
emission permits are used within the country instead of generating export
incomes. Because of this, consumer income and demand fall and therefore
so do the factor prices, resulting in an increased share of the original production being exported. In Sweden and Poland, factor prices fall less in the
case of a regional permit market than with national ones, but for different
reasons. In Poland, income from exporting permits combined with the high
level of abatement contribute to keep the factor prices up. In Sweden, the
chance to buy cheap permits leads to smaller changes in the relative prices,
and thereby smaller reductions in the factor prices. Thus, in Sweden, the
negative effect on the factor prices from an increased import cost is out-
weighed by the fact that the relative prices change less with a regional
permit market, and therefore induce smaller changes in consumption and
export. In Denmark, the factor prices decrease somewhat more in the case
of a regional permit market, reflecting that the increased import of permits
reduces consumer income and demand.

Transfers between countries

The Baltic republics, which incur the highest losses in utility under national
permit markets, are the main winners of a regional permit market. Other for
Poland, all countries gain from switching from a system with a national
market to a regional permit market. As mentioned earlier, the initial distribution of emission permits between the countries is crucial for the outcome
of the cooperation. A relevant question is therefore, is it possible to rearrange
the initial distribution in such a way that all countries gain from coordinating the emission reductions and in that case, how large do the changes
need to be?

The redistribution needed gives Poland a larger initial endowment of
emission permits. The question is whether a large enough amount can be
given to Poland, without taking too much away from some other country. It
turns out that such a solution is not very hard to implement. By decreasing
the initial endowments given to Denmark and the Baltic republic by 5 per
cent, and the Swedish endowment by 7 per cent, we find a solution where
all countries are better off with the regional market, in terms of utility. This
solution is of course not unique, and may seem very arbitrary. The important result is, however, that relatively small changes from the principle of
grandfathering are needed to make coordination of the efforts attractive to
all countries.

Through the larger allocation of permits, it is possible for Poland to
increase the export of permits. This keeps the consumer demand up, and
thereby also the prices, which will increase more than in the previous case.
While the factor prices fell in Poland in the case of grandfathering, they
increase after the redistribution. The level of abatement and production is
the same as before, but more of the original production is consumed within
the country due to the increased consumer income. Due to the lower permit

price, the relative prices change less, and combined with the effect of increased consumer demand, Poland manages to reduce its production of nitrogen-related goods less. Baltic republics incur the highest loss from this redistribution. The lower endowment of nitrogen permits affect the consumer income and demand negatively, and thereby also the factor prices, which fall compared to the case of grandfathering.

In summary, Sweden, Denmark and the Baltic republics lose relative to an allocation according to the grandfathering principle, while Poland gains. Still, all countries are better off than if national permit markets are used. So, with a relatively small modification away from the principle of grandfathering, we have found a solution in which all countries gain from coordinating their emission reductions.

CONCLUSIONS

In this study we have compared two different nutrient-reduction enabling strategies – national markets for tradable emission permits, and one permit market for the region as a whole. The countries included are Sweden, Denmark, Poland and the Baltic republics (Estonia, Latvia and Lithuania). We also experimented with the size of the redistribution of the permit endowments needed to make all countries better off with a regional permit market compared with national ones.

Even if the costs of reducing nitrogen emission are smaller for the region as a whole, when regional trade in permits is allowed, it is not certain that all the countries involved will gain from cooperation. Although Poland receives extra income from exporting permits, this is not enough to compensate for the increased cost of reducing the emissions. The country loses because the negative effect generated by the export income from selling permits outweighs the positive effect generated by the export income from selling permits. Crucial here is the initial distribution of permits between the countries. To make the regional permit market comparable to the national ones, we have used the principle of grandfathering when distributing the permits. So, even though the countries had exactly the same endowment of emission permits in both cases, the opening up of trade between the countries turned out not to be beneficial for one of the countries involved. However, by experimenting with the size of the transfer needed to also make Poland better off with a common market, we found that only a rather small change from the principle of grandfathering is required. By reducing the allocation of permits to Sweden, Denmark and the Baltic republics, and reallocating them to Poland, a solution was found in which all countries lose less as compared with national permit markets.

It should be noted, however, that if the Baltic Sea Declaration is fully implemented, it will be costly and have large consequences on all the economies as shown in Chapter 4. Consumption and production will fall, as a lot of resources are required to abate the emissions. Also, the patterns of consumption and production will change, due to large changes in the relative prices between clean and polluting goods. In terms of utility, the Baltic republics and Poland will lose the most, while the fall in utility is more modest in Sweden and Denmark. So, regardless of whether the targeted reductions are met independently by the countries involved, or by coordinating efforts in the region as a whole, it will be costly to carry them through. Chapter 6 has, on the other hand, presented estimates of the 'environmental gains' that such nutrient-reduction policies could possibly generate.

10 CONCLUSIONS

Ing-Marie Gren, Kerry Turner and Fredrick Wulff

The ultimate purpose of this study has been to identify conditions for an efficient management of nutrient loads to the Baltic Sea. In doing so, we applied a three stage approach. First, we needed to have a good understanding of the current environmental pressures impacting on the Baltic. This implied the creation of a geographical information system (GIS) database which was specifically orientated towards the issue of eutrophication and its causal factors. Then, based on the overall description of the drainage basin, an economic and ecological evaluation of the ministerial agreement for a 50 per cent nutrient-reduction strategy was conducted. The third stage was to simulate the impacts of different strategies for cost sharing between countries, policy coordination and nitrogen permit markets. In carrying out this work we have, we believe, demonstrated that it was possible to conduct an extensive interdisciplinary research project that has generated a range of important empirical and policy-relevant results. One novel feature of the project was the combination of natural science and economic models that allowed for the investigation of a number of realistic environmental response options. The primary achievement of the research was not so much in terms of theoretical advances (in the science or the social science of the project) but in terms of significant empirical findings. For the first time, data on land use, population, costs of nutrient reductions and associated welfare impacts have been analysed on the entire scale of the Baltic Sea and its drainage basin. We will now briefly discuss what we regard as the most significant empirical findings and associated policy implications.

MAIN EMPIRICAL FINDINGS

A great deal of new information on the Baltic region's ecological support systems and land use and population distribution patterns has been generated and organized via a GIS system. The aim was to get an insight into the region-wide environmental pressures that have been building up over time. The part of the drainage basin within the territory of Poland was found to be a key area in terms of nutrient-emission sources. This area contains some 45 per cent of the population of the entire basin and around 40 per cent of intensive arable and livestock areas.

It was also possible to aggregate the seven sub-drainage basin areas into three meaningful general drainage basins – the Bothnian Bay, Bothnian Sea, and the Baltic Proper – at least in the context of nitrogen and phosphorus loads. Based on the GIS data, nitrogen and phosphorus loads from the catchment regions into these three drainage basins were estimated. A novel feature of this calculation was that the loads to the coastal waters could be related to emission sources within the drainage basin. Such a connection is necessary for the accomplishment of a cost-effectiveness analysis of different nutrient-reduction measures. The GIS land use data was used to calculate leaching and load to the Baltic Sea from nutrients deposited on agricultural land and from population centres. The nutrient load calculations highlighted the important role of Poland, with loads of nitrogen and phosphorus accounting for about 33 per cent and 66 per cent of total loading respectively. Germany is another important contributor to the nitrogen deposition in the Baltic Sea, due to its high atmospheric emissions of nitrogen oxides and ammonia. More generally, agriculture and sewage are the most important sources for both nitrogen and phosphorus loads.

A GIS database was also developed to account for the remaining wetlands in the drainage basin. Wetlands provide a range of different functions, including effective nutrient sinks or traps. Therefore, their distribution pattern has important implications for pollution-abatement policy. Overall, the Baltic region has suffered a major loss of wetland ecosystems, especially in the southern parts of the drainage basin. Wetlands now cover only about 8 per cent of the drainage basin. The remaining wetlands are most extensive in the northern regions and are remote from major centres of population and/or intensive agriculture. It is interesting to note that the restoration of wetlands as nitrogen sinks in the southern parts of the drainage basin turns out to be an important potential low-cost abatement option.

The estimation of the cost-effective allocation of nitrogen reductions again served to illustrate the role that Poland could play. Low-cost abatement opportunities in agriculture and sewage treatment in Poland are important for the minimization of the costs of basin-wide nutrient-reduction targets. In fact, cost-effective reductions in the nutrient loads by 50 per cent implies a potential Polish share of about 50 per cent of the total reduction requirement. In addition to measures reducing the use of nitrogen fertilizers, restoration of wetlands in the southern regions turned out to be a cost-effective measure. Another noteworthy result was that, due to the multi-functional aspects of wetlands and other land use measures, total costs can be reduced if both nitrogen and phosphorus reductions are carried out simultaneously. Such a procedure may reduce total cost by about 5 per cent. Another interesting finding is that the costs of phosphorus reduction are about 25 per cent of the costs of corresponding nitrogen reductions, as measured in percentage changes. However, the costs estimates are surrounded by much uncertainty, in particular with respect to nutrient transports within the drainage basins. The costs may be reduced by about 50 per cent for moderate changes in assumptions relating to the nitrogen-retention rates in Poland, Estonia, Latvia and Lithuania.

This research project has allowed the further development of a model that is capable of providing good predictions of the concentrations of nitrogen and phosphorus in the three major basins of the Baltic, given reductions in the nutrient loads from the different drainage basins. This modelling indicates that loads of nitrogen and phosphorus into the Baltic basin have a differential environmental impact on the Baltic Sea, depending on their spatial location. The northern Baltic areas still possess effective nitrogen and phosphorus traps and their inputs have a relatively small effect on the ambient-quality status of the Baltic Sea because of limited transportation into the Baltic Proper. The opposite is the case for the southern Baltic areas.

A concerted attempt was made to estimate the economic benefits of environmental improvements in the Baltic. Some 14 individual pilot and full monetary valuation studies (using contingent valuation and travel-cost methods) were undertaken across three countries, Poland, Sweden and Lithuania. The studies in Poland and Lithuania represent, as far as we are aware, the first deployment of environmental benefit valuation methods in the context of 'transition' economies. Both use and non-use economic (willingness-to-pay) value estimates were derived for marine and coastal resources. The difference between use and non-use value estimates (use values two to four times greater than non-use values) is similar to that found in previous studies of eutrophication reduction – in Chesapeake Bay, US. The basin-wide estimates for total economic value amount to SEK 31,000 million per year. As was noted in Chapter 7, this figure could be regarded as an overestimate. It clearly indicates, nevertheless, that the benefits from a Baltic Sea clean-up of eutrophication are considerable.

However, the total net benefits (benefits minus costs) are highly dependent on which policies are used. When the countries coordinate their policies, the maximum benefits amount to approximately SEK 13,000 million. However, if this policy is changed to national policies where each country maximizes its own net benefits without considering the impact of other countries, the total net benefits are reduced to about SEK 2000 million. The reason for this large difference in net benefits is due to the relatively small domestic impacts from nutrient reductions in countries with large nutrient loads. For example, only about 10 per cent of reductions in Polish nutrient loads generate gains in Poland. The remaining 90 per cent generate gains in other parts of the Baltic. However, under a national policy, Poland only considers its own 10 per cent welfare improvement, implying less nutrient reductions and net benefits than would be the case with a coordinated policy where all welfare improvements are accounted for. The simulation of a permit market for nitrogen indicated similar results – that a regional market generates higher net benefits than national permit markets. Furthermore, it was also shown that, when dispersion effects on the entire economies are considered, the net effects on total production are relatively small, but there may be large structural impacts.

POLICY IMPLICATIONS

At the international level not all global and regional public good problems necessarily lead to inactivity and the loss of environmental capital, as witnessed by the successful international cooperation that has seen significant reductions in sulphur emissions in Europe and in chlorofluorocarbons (CFCs) under the Montreal Protocol globally. Sandler (1998) has concluded that there is a need to better differentiate between global and regional public good problems in order to direct scarce policy-making resources to where they are most effective. In cases where each nation's contribution to the public good, for example, reductions in greenhouse gas emissions and global warming, adds to the overall level of the good (known as 'summation supply technology') the outcome is often under provision. If the benefit derived from an agent's contribution to the public good is less than the provision cost per unit, the dominant strategy is not to provide any of the public good. An adequate provision would then require the intervention of some supranational structure. Nevertheless, not all environmental pollution scenarios are necessarily linked to summation technology cases. In the case of the acid rain problem, for example, a 'weighted sum' technology applies and the 'do nothing' strategy need not dominate. For sulphur emissions a significant share of a given country's public good provision (impact of sulphur emissions reduction) is enjoyed by the country itself. This is because the wind and weather disperses the sulphur over relatively short distances, unlike nitrogen oxide, which is transferred long distances via atmospheric deposition. It is not surprising then that since 1985, significant cutbacks have been achieved in sulphur emissions, which is not the case for nitrogen oxides. The safe storage/containment of very hazardous wastes such as highly radioactive waste would be an example of a 'best-shot' public good supply technology. The overall level of the public good is equated to the largest individual provision level as the nation that makes the innovation determines the public good level of containment for all nations.

Income distribution can have a significant effect on the provision of the public good. Sandler (1998) believes that richer countries are becoming prone to 'foreign-aid fatigue'. If the trend toward income inequality continues, these rich countries may have little choice but to provide global and regional environmental public good, because of the 'best-shot' or 'summation' technologies involved. Rich countries will be forced to provide 'free rides' for the world community, by either subsidizing the poorest countries' provision, or providing it all. Supranational structures will be increasingly required to deal with pending environmental public good crises. Transaction costs can be very high when these structures are in operation and cost-sharing or refund arrangements will have to be deployed to ease enforcement problems and costs. Market transactions such as, for example, pollution emissions trading can prove effective cost-saving devices when linked to the supranational structures.

The findings from our Baltic study confirm these generic international public good problems. The transition to a more sustainable management strategy for the Baltic Sea and related basin area will require intervention policy in order to regulate and manage markets. The response will need to include modification of property rights structures, environmental pollution and resource taxes/charges, tradable permits, securitization schemes, regulations, local community control measures and other devices to change individual and group incentives and to foster levels of trust and moral commitment. A suite of policy instruments will therefore have to be deployed in order to steer technological advances and to mitigate environmental impacts. More flexible national and international institutions and regulations together with so-called market-based incentive instruments will form a combined enabling strategy. There is, therefore, clearly considerable merit in the adoption of a basin-wide approach to pollution-abatement policy in the Baltic and therefore in the implementation of an integrated coastal-zone management strategy. It is clear that the ambient quality of the Baltic Sea is controlled by the coevolution of both biophysical and socio-economic systems throughout the macro-scale drainage basin. Land-use change and population-increase trends (especially since the 1950s) have served to markedly increase the environmental pressure exerted on the Baltic's natural resource base. The waste assimilation function provided by the Baltic's ecosystems is under the most intense pressure.

The natural vulnerability of the area has been amplified by the magnitude and extent of socio-economic activities, impacts and interventions that have become common place since the 1950s. The economic and environmental systems are now sufficiently interrelated so as to be jointly determined. They are now in a process of coevolution. Due to the sheer scale of economic activity the pollution generated is a pervasive problem across the drainage basin and beyond. Localized solutions, for example, at the municipal level, are no longer sufficient international cooperative agreements and actions are required.

There has been a tendency for policy makers and the public to focus most of their attention on nitrogen loads and their impact on the Baltic. Nitrogen-input reduction has become a prime policy target because of the links to blue/green algae episodes and related public perceptions and concerns. In general, nitrogen inputs are more difficult and costly to reduce, but once a reduction has been achieved, the time lag before an ambient state change is experienced is relatively short (up to 10 years). Phosphorous inputs, on the other hand, can be reduced more effectively and cheaply, but there is a long time lag before environmental improvement occurs (up to 25 years). Although there are a range of feasible individual nitrogen-reduction and phosphorus-reduction measures available, our research indicates that the simultaneous reduction of both nitrogen and phosphorus loadings into the Baltic is more environmentally effective as well as cost-effective. The increased deployment of nitrogen-reduction and phosphorus-reduction measures within existing sewage effluent treatment works, combined with coastal wetlands restoration schemes and changes in agricultural practice, would seem to be a particularly cost-effective option set.

The marginal costs of nutrient-reduction measures increase sharply towards the full works treatment end of the spectrum. This finding suggests that the greatest environmental and economic net benefits are to be gained by an abatement policy that is targeted at areas that lack treatment works of an acceptable standard, rather than at making further improvements to treatment facilities that already provide a relatively high standard of effluent treatment. This finding, combined with our findings relating to the importance of the spatial location of nutrient loading, suggests that nutrient reduction measures in the Polish and Russian coastal zone areas would be disproportionately effective. The financing of such measures remains problematic if only 'local' sources of finance are to be deployed. Non-commercial funding from the European Commission and other international agencies, together with bilateral agreements, could play a vital role in the enabling process for an effective and economic Baltic clean-up programme.

Despite the pioneering nature (in the 'transition' economies) of some of the economic-benefits research, it seems to be the case that a cost-effective pollution-abatement strategy roughly equivalent to the 50 per cent nutrients reduction target adopted by the Helsinki Commission, would generate positive net economic benefits (benefits minus costs) or be roughly neutral in cost-benefit terms. Sweden and Denmark would gain net welfare benefits from the implementation of such a policy. In fact, these countries could potentially gain sufficient net benefits to be able to also pay for some clean-up measures in the Baltic republics and Poland, and still remain better off than they were prior to the implementation of such a pollution-abatement policy. The research into the monetary valuation of environmental benefits also indicated that the public's and experts' perception of environmental quality and quality decline are not necessarily the same.

The interdependencies between the countries due to the marine transport of nutrients underscores the need for coordinated policies. If, instead, each country considered its own benefits and costs, only a small total nitrogen reduction would realistically be achieved. The results also indicate few net losers from a coordinated policy. In fact, only Poland would be a net loser. However, the net gains obtained by some other countries would be enough for compensating Poland for its loss. However, a coordinated policy should also rely on efficiency considerations. A policy of uniform pollution reduction targets is neither environmentally nor economically optimal. Rather, what is required is a differentiated approach with abatement measures being concentrated on nutrient loads entering the Baltic Proper from surrounding southern sub-drainage basins. The northern sub-drainage basins possess quite effective nutrient traps and contribute a much smaller proportionate impact on the Baltic's environmental quality state. Some of the countries within whose national jurisdiction these southern sub-basins lie are also the biggest net economic gainers from the abatement strategy.

Nevertheless, the release of some hazardous wastes remains problematic across the basin, and the agricultural sector also poses problems. While the total farming area in the richer countries is much smaller than in other parts of the Basin, the environmental impact is disproportionately large due to the intensive nature of the farming regimes. Intensive agriculture in Denmark and Sweden and its nutrient pollution runoff is a major contributor to Baltic eutrophication. The Danish area of 12,400 km^2 of farmland discharges some 30,000 tons of nitrogen into the Baltic. Agriculture on a less intensive scale in the Vistula River basin, some 166,000 km^2 in Poland, Belarus, Slovakia and the Ukraine, discharges only 110,000 tons (in 1991) of nitrogen per annum. The challenge for Baltic policy makers in the future is clear, however, if the agricultural sectors in Poland etc develop intensive methods similar to those fertilizer/pesticide-dominated regimes commonplace in Denmark and Sweden, the outlook for the mitigation of the effects of eutrophication is poor.

However, although this book has focused on the eutrophication that generates important environmental effects and related losses in economic welfare, there is the danger that too much 'regulatory effort' will be expended on this issue to the detriment of other, equally important (or perhaps even more important), pollution problems. Thus, the release of hazardous substances into the Baltic remains a concern. Less progress has been made generally across the Baltic region when it comes to hazardous-waste-generation minimization and effective treatment before release. Substances which bioaccumulate in the Baltic environment are potentially very serious threats to the long-term health and integrity of the marine and coastal ecosystems. Therefore, it is important to relate the magnitude and significance of the eutrophication problem to other potentially even more crucial environmental problems. Resource scarcity in a general sense is an ever present problem and every investment allocation decision – to clean up nutrient-related pollution, ban the release of contaminants into the marine environment, etc – has an opportunity cost.

REFERENCES

Aerts, R, Wallén, B and Malmer, N (1992) 'Growth-limiting nutrients in sphagnum dominated bogs, subject to low and high atmospheric nitrogen supply', *Journal of Ecology* 80:131–140

Ambio (1990) Special issue: 'Marine Eutrophication', *Ambio* 14(12): 9–14

Ambio (1994) 'Wetlands and lakes as nitrogen traps', *Ambio* 23.

Andréasson-Gren, I-M, Fleischer, S and Stibe, L (1991) 'Can we implement a cost-efficient nitrogen reduction to the Laholm Bay?'; *Ambio* 20:94

Arrow, K, Leamer, E, Portney, P, Randner, R, Schuman, H and Solow, R (1993) 'Report of the NOAA Panel on Contingent Valuation', *Federal Register*, 58(10): 4601–4614

Aselmann, I and Crutzen, P J (1989) 'Global distribution of natural freshwater wetlands and rice paddies, their net primary productivity, seasonality and possible methane emissions', *Journal of Atmospheric Chemistry* 8:307–358

Baker, L A (1992) 'Introduction to nonpoint source pollution in the United States and prospect for wetland use'. *Ecological Engineering* 1:1–26. Baltic Sea Environmental Proceedings, no 48, Helsinki, Finland

Barett, S (1990) 'Global environmental problems', Oxford Review of Economic Policy 8(1): 68–79

Barett, S (1992) 'International environmental agreements as games', In Pethig, R (ed) *Conflicts in managing environmental resources*, Springer-Verlag, New York, pp11–37

Bateman, I J and Willis, K G (eds) (1999) *Valuing Environmental Preferences*, Oxford University Press, Oxford

Bergland, O, Magnussen, K and Navrud, S (1995) 'Benefit Transfer: Testing for Accuracy and Reliability', *NLH Discussion Paper* D-03/1995, Agricultural University of Norway

Bergman, L (1991) 'General equilibrium effects of environmental policy: A CGE-modelling approach', *Environmental and Resource Economics* 1:43–61

Bergman, L (1995), *A Prototype CGE Model for Environmental Policy Analysis*, Stockholm School of Economics, mimeo, April

Bernes, C, (ed) (1980) 'Sweden's marine environment and ecosystems under pressure, *Monitor 1980*', National Swedish Environmental Protection Board Informs, Liber distributions, Stockholm, Sweden, 207 pp

Bertram, G (1992) 'Tradable emission permits in the control of greenhouse gases', *The Journal of Development Studies* 28(3):423–446

Bower, B T and Turner, R K (1998) 'Characterising and analysing benefits from integrated coastal management', *Ocean and Coastal Management* 38:41–66

Brooke, A, Kendrick, D, and Meeraus A, (1992) *GAMS, A User's Guide*, The Scientific Press

Burniaux, J-M, Martin, J P, Nicoletti, G, and Martins, J O (1992), *GREEN, A Multi-sector, Multi-regional General Equilibrium Model for Quantifying the Costs of Curbing CO_2 Emissions: A Technical Manual*, OECD Working papers, no 116, Paris

Chander, P and Tulkens H (1992) 'Theoretical foundations of negotiations and cost sharing in transfrontier pollution problems', *European Economic Review* 36:388–398

Cooper, P F and Findlater, B C, (eds) (1990) *Constructed Wetlands in Water Pollution Control*, Pergamon Press, Oxford

Costanza, R, Sklar, F H and White, M L (1990) 'Modelling coastal landscape dynamics: process based dynamic spatial ecosystem simulation can examine long-term natural changes and human impact', *BioScience* **40**:91–107

Dahlberg, K, Hallin, A K and Wulff, F, (1995) Phosphorus versus nitrogen abatement in the Gulf of Riga – a nutrient budget model approach, Proceeding, International Congress on Modelling and Simulation, November 1995, University of Newcastle, Australia

Dahlberg, K, Hallin, A-K and Wulff F (1996) Phosphorus vs Nitrogen abatement in the Gulf of Riga – a nutrient budget model approach, Department of Systems Ecology, Stockholm University, mimeo

Daily, G C (ed) (1997) *Nature's Services: Societal Dependence on Natural Ecosystems*, Island Press, Washington, DC

Diamond, P A and Hausman, J A (1994), 'Contingent valuation: is some number better than no number?', *Journal of Economic Perspectives*, **8**:45–64 Dordrecht, The Netherlands

Eckerberg, K, Gren, I-M, and Söderqvist T, (1996) *Policies for Combating Water Pollution of the Baltic Sea: Perspectives from Economics and Political Science*, Proceedings from a workshop 27–28 September, Beijer International Institute of Ecological Economics, Royal Swedish Academy of Sciences, Stockholm

Elofsson, K (1997) *Cost-effective reductions in the agricultural load of nitrogen to the Baltic Sea*, Licentiate thesis, Department of Economics, Swedish University of Agricultural Sciences

Environmental Protection Department of the Republic of Lithuania (1992) *National Report to the United Nation Conference on Environment and Development*, Vilnius

Etnier, C and Guterstam, B (eds) (1991) Ecological engineering for wastewater treatment, Proceedings of the international conference, 24–28 March, Stensund Folk College, Bokskogen, Gothenburg, Sweden

Ewel, K and Odum, H T (eds) (1984) *Cypress Swamps*, University Press of Florida, Gainsville

Ewel, K C (1997) 'Water quality improvement by wetlands', *in Nature's Services: Societal Dependence on Natural Ecosystems,* by Daily, G C (ed), Island Press, Washington DC pp 329–344

Fitz, H C, Costanza, R and Reyes, E (1993) *The Everglades Landscape Model (ELM): Summary Report of Task 2, Model Development*, Report to the South Florida Water Management District, Everglades Research Division

Fitz, H C, DeBellevue, E B, Costanza, R, Boumans, R, Maxwell, T, Wainger, L and Sklar, F (in press) Development of a general ecosystem model (GEM) for a range of scales and ecosystems, Ecological Modelling

Fleisher, S, Stibe, L and Leonardsson, L (1991) 'Restoration of wetlands as a mean of reducing nitrogen transport to coastal waters', *Ambio* **20**:271–272

Folke, C, Jansson, Å, Larsson, J, and Costanza, R (1997) 'Ecosystem appropriation by cities', *Ambio* **26**:167–172

Freeman, A M III (1993) *The Measurement of Environmental and Resource Values: Theory and Methods,* Resources for the Future, Washington, DC

Germain, M, Toint, P L and Tulkens H, (1995) 'International negotiations on acid rains in Northern Europe: A discrete time iterative process', in Xepapadeas, S (ed), *Economic Policy for the Environment and Natural Resources*, Edward Elgar, London

Gren, I-M (1995a) 'The value of investing in wetlands for nitrogen abatement', *European Review of Agricultural Economics* **22**:157–172

Gren, I-M (1995b) 'Cost and benefits of restoring wetlands: two Swedish case studies', *Ecological Engineering* **4**:153-162.

Gren, I-M, Folke, C, Turner, R K and Bateman, I (1994) 'Primary and Secondary Values of Wetland Ecosystems', *Environmental and Resource Economics*, **4**:55–74

Gren, I-M and Zylicz, T (1993) *Cost-Effectiveness of the Baltic Sea Clean-Up: Will Wetlands Reconcile Efficiency with Biodiversity?*, Beijer Discussion Paper no 24, Stockholm, The Royal Swedish Academy of Sciences

Gren, I-M, Folke, C, Langaas, S, Sandström, M, Söderqvist, T and Wulff, F (1995) *'Costs and Benefits from Nutrient Reductions to the Baltic Sea'*, The Beijer Institute, The Royal Swedish Academy of Sciences, Stockholm, mimeo.

Gren, I M, Elofsson, K and Jannke, P (1995) *Costs of Nutrient Reductions to the Baltic Sea*, Beijer Discussion Papers Series no 70, Beijer International Institute of Ecological Economics

Gren, I-M, Elofsson, K, and Jannke, P (1998) 'Costs of nutrient reductions to the Baltic Sea', *Environmental and Resource Economics*

Gren, I-M, Jannke, P and Elofsson, K, (1995) *Costs of Nutrient Reductions to the Baltic Sea*, Technical report, Beijer Discussion Papers Series no 70, Beijer International Institute of Ecological Economics, Royal Swedish Academy of Sciences, Stockholm

Gren, I-M, Jannke, P and Elofsson, K, (1997) 'Cost-effective nutrient reductions to the Baltic Sea', *Environmental and Resource Economics* **10**(4):341–362

Gren, I-M, (1998) 'International versus national policies for nitrogen reductions to the Baltic Sea', *Environmental and Resource Economics*, in review

Gren, I-M, and Brännlund, R, (1995) 'Enforcement of regional environmental regulations: nitrogen fertilizers in Sweden', in Hanna, S and Munasinghe, M (eds) *Design Principles of Property Rights System,* The World Bank and the Beijer Institute, Washington, USA and Stockholm, Sweden

Gren, I-M, Söderqvist, T, and Wulf, F (1997) 'Nutrient reductions to the Baltic Sea: ecology and economics', *Journal of Environmental Management* **51**:123–143

Hallin, A-K (1994) *Baltic Drainage Basin Matrix of Nutrient Loads*, Department of Systems Ecology, Stockholm University

Hammer, D A (1989) *Constructed Wetlands for Wastewater Treatment: Municipal, Industrial and Agricultural*, Lewis Publishers, Chelsea, Michigan

Hanemann, W M (1994), 'Valuing the environment through contingent valuation', *Journal of Economic Perspectives* **8**:19–43

Hanley, N and Milne, J (1996) 'Ethical beliefs and behaviour in contingent valuation', *Journal of Environmental Planning and Management* **39**:255–272

HELCOM (1987) '*First Baltic Sea Pollution Load Compilation*', Baltic Sea Environment Proceedings, no **20**, Helsinki, Finland

HELCOM (1993) *The Baltic Sea Joint Comprehensive Environmental Action Programme*, Baltic Sea Environmental Proceedings, no 48, Helsinki, Finland

HELCOM (1993) '*Summaries of the Pre-feasibility Studies*', Baltic Sea Environment Proceedings, no **46**, Helsinki, Finland

HELCOM (1996) *Third Periodic Assessment of the State of the Marine Environment of the Baltic Sea 1989–93; Background Document*, Baltic Sea Proceedings, no 64B, Helsinki, Finland

Hjort, A (1992) *Building International Institutions for Environmental Protection: The Case of Baltic Sea Environmental Cooperation*, Department of Water and Environmental Sciences, Linköping University, Sweden

Hoel, M, (1992) 'International environment conventions: the case of uniform reductions of emissions', *Environmental and Resource Economics* **2**:141-159

Horner, R R (1986) A Review of Wetland Water Quality Functions, paper presented for the Proceedings of the Conference on Wetland Functions, Rehabilitation, and Creation on the Pacific Northwest: The state of our understanding, July 1986, 31 pp

Howard-Williams, C (1985) 'Cycling and retention of nitrogen and phosphorus in wetlands: a theoretical and applied perspective', *Freshwater Biology* **15**:391–431

Howe, C W (1994) 'Taxes versus tradable discharge permits: a review in the light of the US and European experience' *Environmental and Resource Economics* **4**:151–169

Iversen, T Halvorsen, N, Mylona, S and Sandnes, H (1991) Calculated budgets for airborne acidifying components in Europe, 1985, 87, 88, 89, 90. EMEP/ MSC-W Report 1/91, Oslo, Norge

Jacks, G, Joelsson, A and Fleischer, S (1994) 'Nitrogen retention in forest wetlands,' *Ambio* **6**:358–362

Jansson, Å, Folke, C and Langaas, S (1996) 'Quantifying the nutrient retention capacity by wetlands in the large scale drainage basin of the Baltic Sea', Beijer Discussion Papers Series no 69, Beijer International Institute of Ecological Economics

Jansson, Å, Folke, C, and Langaas, S (1998) 'Quantification of the nitrogen retention potential by natural wetlands in the Baltic Sea drainage basin', *Landscape Ecology* **13**: 249–268

Jansson, M, Andersson, R and Berggren, H (1994) 'Wetlands and lakes as nitrogen traps', *Ambio* **6**:320–325

Johanesson, Å and Randås, P (1996) *Economic Impacts of Reducing Nitrogen Emissions into the Baltic Sea*, Paper presented at the EAERE VII conference, Lissabon, June, 1996

Johanneson, Å and Randås, P (1995) *Economic Impacts of Reducing Nitrogen Emissions into the Baltic Sea. The Effects of Implementing Markets for Emission Permits Nationally vs Regionally*, Licentiate thesis, Department of Economics, Stockholm School of Economics, Sweden

Kaitala, V, Mäler, K-M and Tulkens, H (1995) 'The acid rain game as a resource allocation process with an application to the international cooperation among Finland, Russia and Estonia', *Scandinavian Journal of Economics* 97(2):325–343

Klochak, R J (1993) Vegetation planting and monitoring at the Nykvarn and Oxelösund created wetlands, Sweden. Valle Exchange Program, final report. Tema Vatten/SMHI Linköping, Sweden

Kneese, A and Bower, B (1968) *Managing water quality: Economics, technology, institutions*, John Hopkins Press for Resources for the Future, Baltimore

Knight, R L (1992) 'Ancillary benefits and potential problems with use of wetlands for nonpoint source pollution control', *Ecological Engineering* 1:97-113

Kriström, B (1993) 'Comparing Continuous and Discrete Contingent Valuation Questions', *Environmental and Resource Economics* 3:63–71

Krysanova, V, Gornaja, L and Vasilyev, A (1995) '*Baltic Drainage Basin Project. Progress Report from the Estonian Team*', Institute of Economics, Estonian Academy of Sciences, Tallin

Krysanova, V, Meiner, A, Roosaare, J and Vasilyev, A (1989) 'Simulation modeling of the coastal waters pollution from agricultural watershed', *Ecological Modelling*, 49:7–29

Langford, I and Bateman, I (1993) *Welfare Measures for Contingent Valuation Studies: Estimation and Reliability*, CSERGE Working Paper GEC 93–04

Larsson, J, Folke, C and Kautsky, N (1994) 'Ecological limitations and appropriation of ecosystem support by shrimp farming in Colombia', *Environmental Management* 18:663–676

Larsson, U, Elmgren, R and Wulff, F (1985) 'Eutrophication and the Baltic Sea: Causes and Consequences', *Ambio*, 14:9–14

Laurson, P, Melzer, A and Zylicz, T (1995) *A Strategy to Enhance Partnerships in Project Financing for Environmental Investments in Central and Eastern Europe*, Report for the EBRD, London

Leonardsson, L (1994) *Wetlands as Nitrogen sinks, Swedish and International Experience*, Swedish Environmental Agency, Report no 4176, Stockholm, Sweden

Löfgren, S and Olsson, H (1990) *Tillförsel av kväve och fosfor till vattendrag i Sveriges inland, Underlagsrapport till Hav-90, aktionsprogram mot havsföroreningar*, Swedish Environmental Agency, Stockholm, Sweden

Mäler, K-G (1991) 'International environmental problems', in Helm D (ed), *Economic Policy Towards the Environment*, Blackwell, Oxford

Mäler, K-G (1993) The Acid Rain Game II, Beijer Discussion Papers Series no 32, Beijer International Institute of Ecological Economics, Stockholm, Sweden

Malmer, N (1962) Studies on mire vegetation in the Archaean area of south western Gotland, (south Sweden) 1 vegetation and habitat condition on the Akahuit mire. Opera Botanica 7:1–322

Mander, U and Mauring, T (1995) 'Nitrogen and phosphorus retention in natural ecosystems', *Functioning Appraisal of Agricultural Landscape in Europe* (EUROMAB and INTECOL Seminar 1992) pp77–94, edited by Ryszkowski, L and Balazy, S, Research Centre for Agricultural and Forest Environment, Polish Academy of Science Poznan, 1994

Markowska A, and Zylicz, T (1996) *Costing an International Public Good: The Case of the Baltic Sea*. Paper presented at the EAERE VII conference in Lisbon, June, 1996

Ministry of Finance 1992, *The Swedish Budget 1992/93*, Allmänna Förlaget, Stockholm

Mitsch, W J and Gosselink, J G (1993) *Wetlands* (2nd edt), Van Nostrand Reinhold, New York

Mitsch, W J (1992) 'Landscape design and the role of created, restored and natural riparian wetlands in controlling nonpoint source pollution', *Ecological Engineering* 1:27–47

Mitsch, W J and Jörgensen, S E, (ed) (1989) *Ecological Engineering: An Introduction to Ecotechnology*, John Wiley & Sons, New York

Montgomery, W D (1972) 'Market in licenses and efficient pollution control program', *Journal of Economic Theory* 5(3):35–418

Nichols, D S (1983) 'Capacity of natural wetlands to remove nutrients from wastewater', *Journal Water Pollution Control Federation* **55**:495–505

Nord (1993) *The Nordic Environment – Present State, Trends and Threats*, The Nordic Council of Ministers, Nord 1993:12

Nordic Council of Ministers (1994) *Yearbook of Nordic Statistics* 19, Nord 1994:1, vol 32, Copenhagen

Ochrona Œrodowiska (1995) Central Statistical Office, Warsaw [*Environmental Statistics Yearbook*]

OECD data files [OECD]

Olsen, S, Robey, J, Kerry, M (1997) 'A common framework for learning from ICM experience', *Ocean and Coastal Management*, **37**:155–174

Portney, P (1994) 'The contingent valuation debate: why economists should care', *Journal of Economic Perspectives*, **8**:3–17

Rees, W E and Wackernagel, M (1994) 'Ecological footprints and appropriated carrying capacity: measuring the natural capital requirements of the human economy', in Jansson, A-M, Hammer, M, Folke, C, and Costanza, R (eds) *Investing in Natural Capital: The Ecological Economics Approach to Sustainability* Island Press, Washington, DC, pp362–390

Robinson, S, Subramanian, S, and Geoghegan, J A (1994) 'Modelling air pollution abatement in a market-based incentive framework for the Los Angeles basin', in Klaases, G and Försund, F R (eds) *Economic instruments for air pollution control*, Kluwer Academic Publishers, Dordrecht, pp46–67

Rocznik Statystyczny (1995) Central Statistical Office, Warsaw [*Statistical Yearbook 1995*]

Rosenberg, R, Elmgren, R, Fleisher, S, Jonsson, P, Persson, G, and Dahlin, H (1990) 'Marine eutrophication case studies in Sweden', *Ambio* **19**:102–108

Rosswall, T and Granhall, U (1980) 'Nitrogen cycling in a subarctic ombrotrophic mire', *Ecological Bulletin* **30**:209–234

Russell, C and Shogren, J (eds) (1993) *Theory, Modelling, and Experience in the Management of Nonpoint-source Pollution,* Kluwer Academic Publishers. Dordrecht, The Netherlands

Rydlöv, M, Hasslöf, H, Sundblad, K, Robertson, K and Wittgren, H B (1991) Wetlands – Vital Ecosystems for Nature and Societies in the Baltic Sea Region, World Wide Fund for Nature, WWF, report to the Helcom Ad Hoc Level Task Force November, 1991

Sanden, P and Rahm, L (1991) 'Nutrient trends in the Baltic Sea', *Environmetrics* 1993, no 4(1), pp75–103

Sandler, T (1998) Global and Regional Public Goals: A Progress for Collective Action, Fiscal Studies **19**:221–247

Sandnes, H (1993) *Calculated Budgets for Airborne Acidifying Components in Europe,* Meteorological Synthesizing Centre-West, The Norwegian Meteorological Institute. PO Box 43 Blindern, N-0313 Oslo 3, Norway

Sandström, M (1996) *Recreational Benefits from Improved Water Quality: A Random Utility Model of Swedish Sea-Side Recreation*, Licentiate thesis, Department of Economics, Stockholm School of Economics

Slapokas, T (1991) *Influence of Litter Quality and Fertilisation on Microbial Nitrogen Transformations in Short-rotation Forests*, PhD-thesis, Department of microbiology, Swedish University of Agricultural Sciences, Uppsala. Report 49

Small, E (1972) 'Ecological significance of four critical elements in plants of raised *Sphagnum* peat bogs', *Ecology* **53**:498–503

Söderqvist, T (1994) *Baltic Sea Questionnaires: Pilot Work, Phase 1*, Beijer International Institute of Ecological Economics, The Royal Swedish Academy of Sciences

Söderqvist, T (1996a) *Contingent Valuation of a Less Eutrophicated Baltic Sea*, Beijer Discussion Papers Series no 88, Beijer International Institute of Ecological Economics, Stockholm, Sweden

Söderqvist, T (1996b) 'Economic valuation of the environment: methods and the swedish experience', pp5–73 in Ministry of Finance, *Experts' Reports to Skatteväxlingskommittén SOU 1996:117*, Stockholm

Söderqvist, T (1998) 'Why give up money for the Baltic Sea? motives for people's willingness (or reluctance) to pay', *Environmental and Resource Economics* **12**:249–254

Stålnacke, P (1996) *Nutrient Loads to the Baltic Sea,* PhD-thesis, Department of Water State Committee for Statistics of the Republic of Latvia (1994), *Statistical Yearbook of Latvia 1993,* Riga

Statistical Office of Estonia (1993), *Statistical Yearbook/Statisti Aastaraamat 1993,* Tallin, Statistics Sweden and the Swedish National Institute of Economic Research (1994), *SWEEA Swedish Economic and Environmental Accounts,* Stockholm

Statistical Yearbook (1995) Statistics Sweden, Stockholm

Stephan, G, van Nieuwkoop, R, and Wiedmar, T (1992) 'Social incidence and economic costs of carbon limits', *Environmental and Resource Economics* 2:569–591

Stockholm Environment Institute (1990) 'Forward to 1950: policy considerations for the Baltic environment', *Ambio* Special Report no 7, pp21–24, Stockholm

Sweitzer, J, Langaas S and Folke C (1995) 'Land Use and Population Density in the Baltic Sea Drainage Basin: A GIS Database', *Beijer Discussion Papers* no 63, The Beijer Institute, The Royal Swedish Academy of Sciences, Stockholm

Sweitzer, J, Langaas, S and Folke, C, (1996) 'Land use and population density in the Baltic Sea drainage basin', *Ambio* 25:191–198

Sweitzer, J, Langaas, S and Folke, C (1996) 'Land use and population density in the Baltic Sea drainage basin', *Ambio* 25:191–198, http://www.grida.no/baltic

Tietenberg, T (1995) 'Design lessons from existing air pollution control systems: The United States', in Hanna, S and Munasinghe, M *Property Rights in a Social and Ecologica Context: Case Studies and Design Applications,* The International Bank for Reconstruction and Development/World Bank, Washington, USA

Tilton, D L (1977) 'Seasonal growth and foliar nutrients of *Larix laricina* three wetland ecosystems', *Canadian Journal of Botanics* 55:1291–1298

Torell, L (1997) 'Pollutants from Swedish municipal and industrial outlets into the Baltic Sea', *Ambio Special report* 5:213–218

Turner, R K (1999) 'The place of economic values in environmental valuation', in Bateman I J and Willis, K G (eds) *Valuing Environmental Preferences,* Oxford University Press, Oxford, pp17–41

Turner, R K et al (1995) *The Baltic Drainage Basin Report: EV5V-CT-92-0183,* Final Report of the Project, Norwich, University of East Anglia

Turner, R K et al (1998) 'Coastal management for sustainable development: analysing environmental and socio-economic changes on the UK coast', *Geographical Journal* 164:269–281

Turner, R K et al (1999) 'Managing nutrient fluxes and pollution in the Baltic: an interdisciplinary simulation study', *Ecological Economics,* 30: 335–352

Turner, R K and Pearce, D W (1993) 'Sustainable economic development, economics and ethics', in Barbier, E B (ed) *Economics and Ecology,* Chapman and Hall, London, pp178–194

Verry, E S and Timmons, D R (1982) 'Waterborne nutrient flow through an upland-peatland watershed in Minnesota', *Ecology* 63:1456–1467

Whalley, J and Whiggle, R (1991) 'Cutting CO_2 emissions: the effects of alternative policy approaches', *The Energy Journal* 12(1):109–124

Wulff, F and Niemi, A (1992) 'Priorities for the restoration of the Baltic Sea: a scientific perspective', *Ambio* 21(2):193–195

Wulff, F, Stigebrandt, A and Rahm, L (1990) 'Nutrient Dynamics of the Baltic Sea', *Ambio* 3:126-133

Yurkovskis, A, Wulff, F, Rahm, L, Andrulewisz, A and Rodeigues-Medina, M (1989) 'A nutrient budget of the Gulf of Riga, Baltic Sea Estuarine', *Coastal and Shelf Science* 37:113-127

Zylicz, T (1993) 'The ecological economics of the Baltic Sea', *European Review,* 1(4):329–335

Zylicz, T (1994) *A Survey of the Cost-effectiveness of Investment Projects Co-financed by the Polish National Fund for Environmental Protection,* mimeo, Warsaw University

Zylicz, T, Bateman, I, Georgiou, S, Markowska, A, Dzięgielewska, D, Turner, R K, Graham, A and Langford, I (1995) *Contingent Valuation of Eutrophication Damage in the Baltic Sea Region,* CSERGE Working Paper GEC 95–03, Norwich, University of East Anglia

INDEX